Nakayama, Shigeru, 1928-
Academic and scientific traditions in China, Japan, and the West

DATE			

Academic and Scientific Traditions
in China, Japan, and the West

Academic and Scientific Traditions in China, Japan, and the West

Shigeru Nakayama

translated by Jerry Dusenbury

UNIVERSITY OF TOKYO PRESS

Publication of this volume was assisted by a grant from The Japan Foundation. The translation was supported by a grant-in-aid from the Ministry of Education, Science and Culture, Japan.

This book is a translation of *Rekishi toshite no gakumon,* by Nakayama Shigeru, originally published by Chuokoron-sha, Inc. (© Chuokoron-sha, 1974).

ISBN 4-13-068107-9/UTP 69078
ISBN 0-86008-339-X

Printed in Japan

Contents

Preface

A particular mode of learning does not simply appear. It does not burst forth in the heavens like a solitary celestial nova, only to vanish abruptly of its own accord. Rather, it comes into being only in the context of communication between those who write and those who read, between those who employ the learned idiom to record their observations and those who find such records interesting. The species of learning known today as "science" is an excellent example of this proposition, for the work of the contemporary scientist presupposes a whole complex of ideas, instruments, and institutions without which "scientific activity" is scarcely conceivable. Indeed, modern science might be simply described as the dominant contemporary form of communicable knowledge. The ongoing communication of this knowledge constitutes an academic tradition, and it is with the study of such academic traditions that the history of science as I understand it is properly concerned.

Thus defined, the history of science is constrained by the very nature of its subject matter to regard knowledge itself as a social phenomenon. I say that it is constrained to do so, not because I wish to view particular intellectual achievements as the products of particular societies, but because the particular modes of learning in which knowledge is fabricated as communication constitute the social medium in which it is fashioned, through which it is developed, and by which it is transmitted among people commonly engaged in serious intellectual inquiry. On the basis of this observation, I shall concentrate in this work on a structural analysis of scientific groups.

I am, of course, by no means the first to treat science in social terms. Prominent among my predecessors are a generation of

Marxist historians of science. In a "Social History of Science" project initiated in the 1930s, they focused their attention on the immediate relationship between science and the "mode of production" and sought to demonstrate that science, though apparently an abstract entity quite removed from the entanglements of social life, actually had roots in society.[1] In their terminology, they sought to explain science as a superstructural phenomenon with a social foundation. Partly as a result of these scholars' efforts, the notion of science as a "social product" has now come to be generally accepted. Nevertheless, and despite the fact that the simplicity and clear-cut character of their monistic explanations of scientific development still give them a certain attractiveness, the sterility of the historical materialist history of science project is now quite evident. How is this to be explained?

One of the project's most important limitations was the failure of its leading protagonists to follow the initial propagation of their position with the articulate program of inquiry that might conceivably have enabled this perspective to establish itself at the research level. In the absence of such a program, their studies did not give rise to a new generation of scholars committed to refining their observations and building upon their initiatives. Thus historical materialist studies in the history of science appeared increasingly vulnerable to critics in the orthodox wing of the profession, who were wont to find its narrative crude and overdrawn.

Any attempt to predicate science history solely upon direct correspondence between science and the mode of production soon runs into serious difficulties. This scheme has, for instance, been employed to explain progress in pure science during a given period as a consequence of increased economic productivity. Yet it is equally conceivable that heightened productivity might encourage large numbers of potential scientists to flock to the applied

[1] As a scholarly movement this tradition was launched by B. M. Hessen's important paper, "The Social and Economic Roots of Newton's *Principia*," delivered at the second meeting of the International History of Science Association in 1931. J. D. Bernal's *Science in History* (1954) is usually regarded as its most complete statement. In Japan there is a social history of science tradition that antedates Hessen's paper by two years, having originated with the publication of Ogura Kinnosuke's study, "Kaikyū Shakai no Sanjutsu" ["The Arithmetic of a Class Society"] (1929). The English reader can get some feel for this tradition by reading *Science and Society in Modern Japan* (Nakayama, Swain, and Yagi, 1974).

fields, leaving the vineyards of pure science largely untended. This seems in fact to be what happened to English chemistry during the first half of the 19th century. Thus the introduction of a middle factor into the equation could well lead to precisely the opposite conclusion. On the other hand, German chemists showed little interest in the expanding 19th-century world of business and industry, holding fast to a pattern of pure academic research that had produced major achievements even prior to the industrial revolution.

Figure P-1 represents one attempt to construct a more complex analytical model. In this model, the core of the history of science is occupied by the development of scientific theory. This history is given to us in the form of chronological tables and the annals of inventions and discoveries. Immediately surrounding this innermost core is the world of scientific thought out of which theory emerges. Orthodox historians of science have recently produced a significant body of research in this area. Beyond this "intellectual history of science" there is the immediate professional environment in which the scientist carries on his daily activities—the research group or groups with whom he works, the academic associations to which he belongs, the university or research institute with

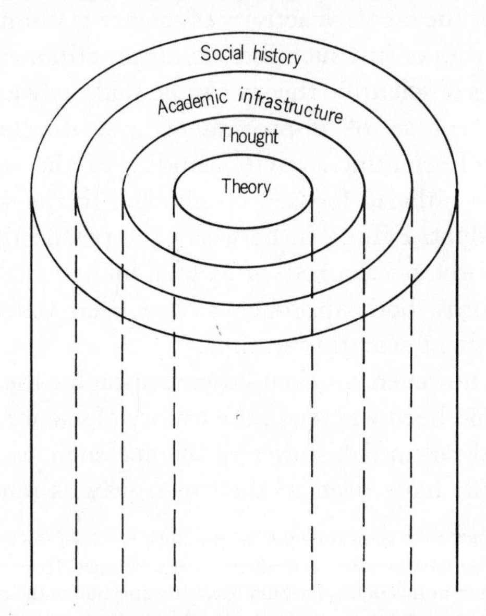

Figure P–1

which he is associated. This I call the academic infrastructure. Finally, in the outermost circle, we have the larger society.

This model illustrates one of the sources of the loose reasoning that has often characterized studies written from the historical materialist point of view: the tendency to juxtapose and directly link the theoretical developments represented at the core of the model with the larger society on the outer rim, ignoring the intermediate elements. It is my contention that a convincing account of the growth and development of science cannot be written unless a serious effort is made to fill in this intermediate realm. Much more extensive inquiry into what I have called the academic infrastructure is particularly urgent. To this end, I have consciously attempted in these pages to reverse the common pattern and proceed outward gradually from the learning nucleus to the general societal setting.[2] Such an approach has considerable value, at least in supplementing the work of those who have viewed the social base as determinative. For even if one adopts this basic explanatory scheme, he is not thereby absolved of the task of linking outer rim with inner core in an intelligible manner. Indeed he cannot ignore this obligation without seriously impairing the authenticity of his model—at least in the eyes of the practicing scientist. For the creative activity of science is ultimately reducible to the activities of just such individual practitioners.

If histories of scientific theory can be said, by way of analogy, to record the course of improvement and development in the "product" of scientific activity, studies in the social history of science have hitherto focused on changes in the social "market" for this product. Falling in between, the perspective of the "producer" has been overlooked, or at least inadequately appreciated. In other words, both approaches have been wanting in a structural analysis of scientific groups.

There is, however, an even larger reason for the arid quality of much that has been written in the history of science. General introductory and cultural histories of science such as that of George Sarton (1948) have been in their own way as removed from the

[2] In this sense I have concentrated on the links between the growth and development of academic theories and the academic infrastructure. If one can think in terms of a division of labor at this point, I would leave the question of the relationship between the institutional structures of scholarly life and socioeconomic history to the general social historian.

actualities of history as historical materialist studies. Singularly macroscopic in outlook, all have surveyed the development of science and society and linked them together in a variety of ways. But in so doing, they have treated science as a structure to be contemplated rather than as a dynamic historical activity involving creation, development, and transmission. To view human intellectual activity only in terms of its output is like tracing the movements of living creatures by their excrement. Just as the history of art is commonly written not as a parade of art objects but as the story of artists and their ideas, so the history of science must come to treat science as an activity and not simply as a series of products on display in a showroom.

There are any number of books that take the system of science as an established edifice and proceed thereupon to advance a general theory of science and its development. But I am loath to engage in the type of discussion that seems in effect to be centrally concerned with the labeling of preserved specimens. Unless we think of science as something that is alive and changing, its shifts and turns will continually confound our best efforts at interpretation, and we shall forever be unable to keep pace with its development. For in the very act of arresting a growing, active tradition of learning and attempting to order it systematically, one misrepresents creative activity and embraces a distortion.

The view that science is something historically quite specific and well-defined likewise involves a contemplative approach to intellectual activity. Thus the notion that science is Greek science and all those modes of learning that do not conform to the Greek pattern are not science would view all such activity in terms of and through its perception of the Greek "scientific edifice." When intellectual activity is understood in terms of the acts of creation, development, and transmission, however, it emerges as a human actvity that could and in fact has occurred everywhere. Thus it becomes possible to speak of Chinese science, Indian science, and even Hottentot science.[3]

[3] In this book I have avoided any hard and fast definition of science. From my point of view, it is not very meaningful to take the science of a particular period or region and make it into a kind of ideal type. In fact, there is a group of us who no longer feel we are speaking properly unless we preface the word science with an adjective: Chinese science, 19th-century science, etc. For a brief discussion of the variety of meanings science has had in modern times, see Chapter 6.

This book is consistently informed by the view that no mode of learning has its reason for being given to it *a priori*. When one surveys the historical rise and fall of particular modes of learning, the suprahistorical significance that we so readily attach to them as fundamental ways of knowing and perceiving simply melts away. One need only recall how, at a certain point in time, the ancient Scholastic and Confucian traditions were suddenly abandoned, how what had once seemed so immutable came to be largely ignored by subsequent generations. Nor does history want for instances in which modes of learning that flourished in one cultural area went completely uncultivated in others. If such observations lead to the relativization of all that appears fixed and permanent, this is, after all, one of the functions of historical research.

Academic traditions in general and the history of scientific inquiry in particular are successions of ordinary acts in which men and women who have been drawn to the work of those who preceded them select, reject, and develop their achievements. It is at this juncture and in the context of this encounter between successor and predecessor that larger historical questions, such as why development followed a particular course and whether there were not other alternatives, properly arise.

In the orthogenetic view that the growth of knowledge runs continuously in a pre-ordained direction along a single track laid across the expanse of time from the beginning of human history, the thrill of being seized by an idea, the adventure of choice, and the trials of doubt and disappointment are transformed into purely mechanical operations. This view squeezes the life from human activity no less than the view that regards the system of science as given. In actuality, at every point in history choices are made that shape and direct the course of learning.

It is easy to fall prey to the deeply entrenched but nonetheless illusory notion that modern science developed along a logical, necessary course from ancient Greece through the Renaissance to the present. In venturing upon this comparative study of East and West, I hope to inject some flexibility into this view through a consideration of alternative patterns of development and to capture, insofar as possible, the dynamic and malleable character of academic traditions.

By the West, I mean in effect the historical mainstream of scientific development. Though there were frequent shifts in its historical center—from ancient Babylonia and classical Greece to the larger world of Hellenistic civilization (extending as far as India), from the Hellenistic world to the Islamic cultural sphere, and from there to Renaissance Italy, 17th-century England, 18th-century France, 19th-century Germany and 20th-century America—this is the stream that eventually grew into contemporary science. By the East, I mean science as it was practiced in the Chinese cultural sphere, in almost complete isolation from this mainstream.

Comparative judgments concerning the academic traditions of East and West tend to be deeply felt and strongly held. Modernists have generally looked at the East in terms of evaluative criteria supplied by the West. In contrast, scholars like Joseph Needham have made every effort to break down Western prejudice by portraying Chinese achievements in the best possible light. The Chinese themselves have often seemed caught between their own cultural tradition and the complex of feelings toward the West typically associated with the May Fourth Movement intellectuals. In this respect, the Japanese occupy a highly strategic vantage point, one that potentially affords them a wide angle of vision. Progenitors of neither tradition, they have little cause to approach the matter with either excessive pride or the sense of inferiority that pride often conceals. Moreover, since the initial influx of Western knowledge during the Edo period and the subsequent Meiji shift to Western modes of learning were preceded by centuries of training in the achievements of Chinese civilization, the course of their intellectual history has long since placed Japanese in a position to compare the two traditions. Japan's academic tradition has produced little that is original; its influence abroad has not been significant. But the Japanese have habitually cultivated the faculty of critical choice. It is finally this sense of my own tradition that has led me to undertake herein the awesome task of comparing the two streams from which it developed.

For better or worse, the history of science is still less than universally accepted and taught in Japanese universities, but during the 1950s I had the good fortune to be trained in this discipline at Harvard University. Here, in the midst of what was as academic an approach to the subject as could be found at the time, I faithfully observed the conventions of normal science and submitted a piece of research that adhered to the paths of scholarship opened by those who had gone before. Even then, however, I had the idea of doing something that went beyond the confines of this kind of activity.

The history of science at Harvard in the 1950s was under the influence of Alexandre Koyré and his brand of intellectual history. (If Marxist studies of the socioeconomic substructure of science can be called "external history," Koyré's work was classic "internal history.") Its popularity can no doubt be understood in a variety of ways, but the threat to the autonomy of intellectual life posed by the rampant McCarthyism of that time may well have given its emphasis on internal history particular significance. Koyré's approach also proved attractive to science historians because it provided them with a platform from which to argue for the legitimacy of their discipline as an independent profession. Armed only with "external history," they would have found it much more difficult to distinguish themselves from institutional or economic historians and to carve out a separate place for themselves in the university.

Like many other Japanese of the generation which came to intellectual maturity in the years after Japan's defeat in World War II, I had been baptized into a science history grounded in historical materialism. This historical materialism had emerged from the war years with a new and commanding authority, and though I was critical where I found it dogmatic, I still thought and worked within the framework it provided. To someone with this background, Koyré-style science history was new and even refreshing. Yet whether because of my background or some other factor, I was never able to feel fully at home with the "internal history" approach. Though I ceased to think in terms of Marxian formulas as such, my desire to deal with the evident links between science and its social foundations persisted. It was at this point that I met Thomas Kuhn.

Kuhn had been strongly influenced by Koyré, but he had not abandoned altogether the attempt to link the "internal" and "external" dimensions of science. *The Structure of Scientific Revolutions* (1962 and 1970) gave expression to his ideas about the nature of scientific activity. Taking the paradigm concept as its point of departure, this work opened up an avenue of inquiry that pointed to the question of the structure of scientific groups. The next logical step was an inquiry into what Kuhn called "the community structure of science," and it would have been quite natural for him to have undertaken this task himself. Having become absorbed in a dispute with philosophers over the paradigm concept, however, he has found it impossible to extend himself in this direction. Thus one of my purposes in writing this book has been to carry Kuhn's pivotal concept into an area where its implications have not yet been fully worked out—that is, into the "external," social history of science.

The paradigm concept is admittedly ambiguous. In one sense, this ambiguity is directly related to its inception as an attempt to give name to an experienced reality inadequately expressed by terminology previously available for describing what the scientist does. If I have ignored the problem of further definition, however, it is because I know from my own experience that however one describes what moves and sustains the scientist in his work, formal theories and factual verification are less important than the comparatively vague and suggestive "something" to which the "paradigm" seeks to point. Should the term be broken down into several others of lesser scope, there is a real danger that the purposes for which the word has been used will be frustrated. Moreover, I am convinced that anyone with any research experience will be able to grasp what is meant by the term.

If, in bringing this experience to consciousness and providing language with which to explore it, the paradigm concept has made a contribution to our understanding of science as an intellectual activity, it has also suggested a new approach to science as a social phenomenon. It has made it possible to think in terms of an analysis that begins with scientific thought and moves outward toward society, thereby reversing the old socioeconomic analysis which began with the social base. The study of science as a social phenomenon is, of course, not the only approach to the history of scientific development. There is and probably always will be the pure "internal history" approach, the history of ideas generated in the human brain—the history of mathematics, for instance. Still, thought commits itself to institutions and social life in such a way that it is only as we are able to depict the history of the interaction between thought and institutions that intellectual history is brought down to earth and comes to have reality for us.

Another thing this book has attempted to do is to set up a discussion of the academic traditions of East and West seen as multiple lines of development. As an East Asian, I cannot be persuaded that the course of scientific development followed by the West is the only logically necessary course. The old Enlightenment view of history took the present time (P) (see Figure P-2) as its point of departure and considered significant only those past events that the contemporary vantage point could identify as being positively linked to the growth of science as currently constituted. Other possible lines of development were simply thrown away. Science historians who work in the Koyré style, on the other hand, chart scientific advance from the formative moment (O) in the history of a given paradigm or mode of learning. When this is done, the line OP no longer stands alone. Other possible courses of development such as OP′ or OP″ come into view. It has been my intention to expand this perspective geographically—to apply it, that is, to the Chinese as well as the Western tradition. To do so is not only to plot the development of Chinese science in terms of its own formative moment (O) (including lines most of which proved abortive), but to construct a frame of reference within which Western scholarship might conceivably not appear to be making progress.

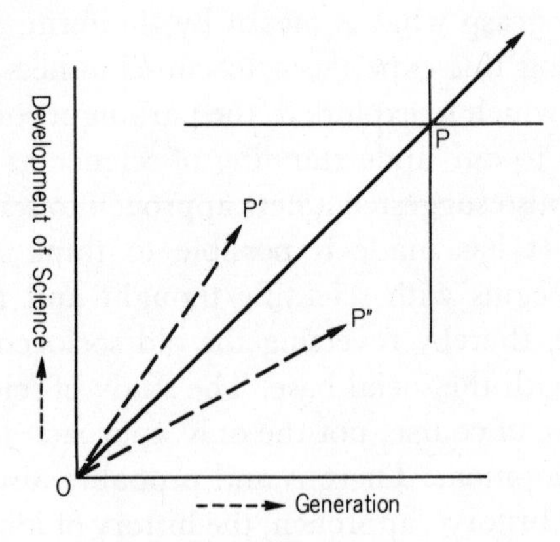

Figure P–2

If the possibility of different criteria of progress and different lines of development is acknowledged, then the question of which standards of judgment to employ and which course to pursue becomes a matter of choice. Faced with a pluralistic situation, it may be tempting to minimize the consequences of choice in the belief that serious application to the task at hand will eventually lead us to one and the same goal regardless of the line we take. This kind of easy optimism is, however, untenable. The view that there is a kind of pre-established harmonious goal, an absolute truth toward which mankind gradually ascends through trial and error—the view espoused by so many and embraced by Lenin in *Materialism and Empirio-Criticism*—is entirely false. Minute by minute, hour by hour, people are at work revising their goals, while from time to time a scientific revolution brings about a reorientation of an entire tradition. Normal science activity resumes after these shifts, but as it proceeds, science advances toward goals increasingly divergent from those that were previously being pursued. Thus the direction and goals of science are the responsibility of each new generation: they are not given *a priori* by some Divine decree. The choices people make are important. Indeed, in its larger contours the history of science might well be seen as a history of choices. In situations of choice, group and societal factors come into play, shaping both the alternatives available and the decisions made. It is for this reason that we are obliged to treat the history of science and learning within the context of scholarly groups and of the general social environment.

My decision to go into science history ultimately goes back to my first simple doubts about the claims, pretensions, and practices of the academic society into which I was being initiated. As I looked around, I found myself in a world of closed circles within which paradigms from across the sea and the precepts of predecessors were meekly followed, a world in which the individual who happened to be uncomfortable with what he was taught was supposed to assume that the problem lay with him and move quickly to accommodate himself to the prevailing view—or, if he were unable to do so, find himself branded "a professional failure."

Did one have any right to expect intellectual creativity under these conditions? Was this really the only kind of atmosphere Japanese scholarship could produce?

There was a long period after World War II when the Japanese academic community was acutely aware of the need to catch up with the United States. In this frame of mind scholars spent their days gleaning what they could from American fields, believing that anything that could not be done in the United States was out of the question in Japan. And though the nation can at long last count itself among the most advanced civilizations in the world, we still have not fully outgrown our old habits. Bewildered by our new situation, we have continued to look abroad for goals, only to be further disconcerted by the confusion and uncertainty we see in the American "main house." At the same time, we do not yet have the confidence to set up our own goals and establish our own lines of scientific inquiry. Our predicament is further complicated by the fact that we live at a time when the image of modern science has begun to lose some of its luster and the implicit value of scientific progress no longer goes unquestioned.

Yet as we are liberated from the goal of "catching up" that has stood before us for a century and begin to look out on new horizons, I feel that there is bound to be a renewed concern with the question of the ends of scholarship and scholarly life, particularly among the younger members of the community. The reader may wonder at the temerity of a professional historian who writes with such reckless disregard for the canons of his discipline. While I have done so in the belief that I have a rather unusual perspective to offer on many of the issues raised in this book, I am under no illusions that I have provided many original answers. In fact, all I have done is to throw out a lot of questions. There is a sense in which any attempt to do more would have been inappropriate, for though I have allowed myself some fairly critical remarks about the scientific profession, I am in no position to be prescriptive. The science historian can provide materials for thinking about the course of scholarly development, but it is the citizens of tomorrow—and particularly the young men and women now embarking on careers in scientific research—who must change it. If this book should serve to stimulate a general and critical discussion of the goals and direction of modern scholarship, I shall be more than happy.

In conclusion I would like to thank Mase Mitsuhiko of the *Chūō Koron* and Satō Shōsuke of Tohoku University who read the manuscript and recommended it for publication.

August 1974 NAKAYAMA Shigeru

Preface to the English Edition

This book is the English translation of *Rekishi toshite no gakumon* (Science and learning as history) (Tokyo: Chuo Koron, 1974). The original Japanese edition is of course addressed to contemporary Japanese readers. In putting it into an English-language context, there are two distinctly different ways of presentation. One is to adhere faithfully to the original text and try to keep the Japanese flavor as much as possible; the other is to edit heavily in order to adapt the text to readers from other cultures with their different backgrounds and knowledge. While I was advised to take the latter course by several readers, the translator favored, and followed, the former, and in the course of thinking about the problem, I came to agree that presenting the text much as it appeared in Japanese was of value.

When I have written in English about Japanese science, I have usually done so by writing—and, insofar as possible, thinking—directly in English rather than following the more circuitous route of writing in my native Japanese first and translating it into English. In doing so, I feel that I am addressing an international audience straightforwardly. This procedure does, however, involve a shortcut, in that it consciously or subconsciously avoids culture-specific concepts and thoughts, sacrificing them to smooth readability. On the other hand, when I translate a Western work into Japanese, I never ask the author to rewrite it to address a Japanese audience. Instead, I (like other translators doing similar work) try to understand and render the Western cultural context as faithfully as possible, often with the aid of interpretive notes.

Thus, the intellectual and cultural traffic seems one-way. It leads to the problem of "English-language imperialism," to which all non-native English speakers are subject. For them, two sets of academic traditions coexist without much integration: the international scientific community, where English is the common language, and a vernacular intellectual tradition for the communication of information and informal discussion. Most scholars live in such a "bilingual" atmosphere; the weight of commitment to each tradition can be arranged on a spectrum with the internationally-oriented physical sciences at one end and the vernacular humanistic studies at the other.

Each tradition has its own infrastructure, merit system, citation group, and audience, without much prospect of integration in the near future. Whether this bilingual intellectual exercise is stimulating or merely inhibits the generation of creative ideas is a subject for future analysis in terms of the sociology of science.

This translation has not been revised to any significant degree for an international audience. Rather, I felt the international audience should be exposed to such Japanese problematics as occur in it, especially in Chapter 6. (Readers knowledgeable about Chinese and Japanese history may find the original Japanese-language version helpful for ideographic renderings of names and works of Asian scholarship.) Perhaps this is one step toward two-way intellectual traffic.

Jerry Dusenbury, a student of Japanese history and language at the time he translated the book, brought a broad knowledge of the subject as well as his considerable translation skills to the project. Nathan Sivin, historian of Chinese science at the University of Pennsylvania, was kind enough to give the entire manuscript a critical reading and suggested a number of useful changes. David Swain, my longtime collaborator in Tokyo, also read and commented on major portions of the text. Susan Schmidt of the University of Tokyo Press ably saw the manuscript through publication. I wish to thank all of them for their contributions to the publication of this volume.

April 1984 Shigeru NAKAYAMA

Academic and Scientific Traditions
in China, Japan, and the West

Chapter 1

Two Styles of Learning

One cannot examine the academic traditions of East and West in the earliest stages of their development without being struck by the remarkable parallels between them. Though this awareness has received a variety of expressions, writings on the subject can be roughly divided into two types: those that seek to compare Babylonia and archaic China and those that would compare and contrast classical Greece with the China of the Warring States period. Behind this duality of focus it is possible to see the presence, in both traditions, of two distinct styles of scholarly activity, styles ultimately grounded in the two general mediums by which human knowledge is communicated. The first, which I shall call the documentary, is centered around the keeping and ordering of written records. The other is characteristically cultivated and transmitted orally in and through polemical discourse and may be conveniently termed the rhetorical.

Documentary Scholarship

The notion that there was cultural contact between the civilizations of Babylonia and ancient China (chiefly in the form of the spread of Babylonian civilization toward China) has been the subject of a long-standing and as yet unsettled debate among Orientalists.[1] But whether or not there was contact, that there are many points of apparent similarity cannot be denied. Foremost among them are records of extraordinary occurrences in the

[1] In prewar Japan this question was the source of a celebrated controversy between Shinjō Shinzō, who claimed that Chinese astronomy represented an original development, and Iijima Tadao, who argued that it had Babylonian roots. See Shinjō Shinzō (1928).

heavens. Abundant both in ancient Chinese documents and on the cuneiform tablets of approximately the 7th century B.C. from the archives of King Ashurbanipal, these records evince an amazing likeness of purpose, method, and function in the activity that gave rise to them. In Babylonia as well as in China, people recorded changes in the heavens in order to ascertain their influence on earthly events. They were engaged, that is, in varieties of astrology.

The implicit suggestion that a discussion of documentary learning might properly begin with an account of ancient astrology may well meet with resistance. The concept itself more readily calls to mind a chronological historical narrative such as the *Spring and Autumn Annals*. Astrology, on the other hand, conjures up images of contemporary fortune-telling arbitrarily manufacturing explanations from dubious data. But the issue is not the interpretation of events, but, rather, the records kept for this purpose. Moreover, in ancient times there was virtually no distinction between the historian and the astrologer. This is particularly evident in China where the term *shih* (史) comprehended both pursuits. Ssu-ma Ch'ien, the most famous of the Chinese astrologer-historians, is best known as an official historian charged with the compilation of court documents, but in his post as Grand Recorder (*t'ai shih ling*) he was also responsible for maintaining astrological records. In other words, "astrology" was originally regarded as historical, documentary science. If one thinks of learning as beginning with the oral traditions of the storyteller, traditions later written down, then these early records of strange and dramatic phenomena in the heavens may simply document what were perhaps the most impressive events recounted by the bard.

Information can be conveyed through written records only if readers are able to apprehend it more or less as the chronicler intended. In this imperative, one can recognize the first beginnings of a conscious orientation toward "objectivity." In recording ordinary historical events there is, of course, ample room for subjective interpretations. But the self-evident character of most events typically documented for astrological purposes (solar eclipses, for example) makes such records resistant to the biases of the transmitter-recorder. It is because they constitute an example of intended "pure objectivity" that I have chosen to look at astrological records as exemplary of scholarly activity in the documentary style.

But can one find in ancient astrology the search for lawful regularities that moderns have come to see as the goal of scientific activity? Both Babylonian and Chinese astrology begin with a recording of the appearance of extraordinary celestial phenomena—solar eclipses, comets, meteors, and the like. Should the outbreak of an earthly calamity such as war or famine coincide with one of these phenomena, then the two are seen as related in some way. For example, let us assume that on several occasions when the sun is in a certain position, a flood occurs in a particular region. The next time a solar eclipse occurs with the sun in the same position, it is taken as an omen of another flood in that region. Forecasts of the coming disaster are made and precautions taken. To those engaged in this activity, the mechanism that linked heavenly phenomena and earthly events remained unknown. But the whole enterprise rested on the conviction that there was a relationship between the two. It was in order to find out all they could about this relationship that people sought to assemble as much data as possible concerning extraordinary heavenly phenomena and natural disasters and to hypothesize about their relations.

Although speculative dogmas (such as the theory of Yin-yang and the Five Elements) were used in forging the interpretative link between occurrences in the heavens and events on earth, the records themselves have outlived the half-baked theories. Just as administrators and members of the legal profession tend to endow precedent with intrinsic significance, so the ancient astrologer regarded past records as an independent and virtually absolute source of authority. Every effort was made to resolve the discrepancy between the records and the new phenomenon within the established system of explanations, usually by an expanded interpretation of new and old data that made it possible to deal with both under the same classificatory rubric. Only unprecedented anomalies generated enough tension to require new interpretation. As a result, revolutionary theories were almost unheard of as the records continued to increase year after year.

Contemporary readers will not find this surprising. Believing with modern science that links between extraordinary celestial occurrences and terrestrial disasters are either nonexistent or irrelevant, they do not share the ancients' hope that the faithful accumulation of such data will eventually make possible the extrapolation of "empirical laws" governing this relationship. They are much more likely to be amazed that the ancients continued to pore through stack upon stack of old documents without becoming fed up with it all. Yet the discipline of the documentary style is readily recognizable. Although it does not necessarily call for a creative mind, it demands the same painstaking care, the same devotion to clerical routine, as bookkeeping.

The clerical collection and classification of documents also requires a pool of industrious, full-time workers, a condition that can be met only by a highly developed bureaucracy. In this sense, this style of scholarship seems particularly well suited to Chinese political culture. Its emergence under the "Oriental despotisms" of Babylonia and China is also suggestive. For unlike the horoscope astrology of today, the astrology of extraordinary occurrences in the heavens was not concerned with the fate of individuals as such, but with the divining of answers to questions of concern to the monarch as absolute ruler of the state (Nakayama, 1964). From the dawn of recorded history down to modern times, the major task of the Chinese observatory was to make and interpret the types of astrological observations that were deemed relevant to this task. This is the "astronomy" (*t'ien-wen*) so evident in the official dynastic histories of Imperial China.

Rhetorical Learning

The ancient world also witnessed the emergence of another style of scholarly activity, one that stands in sharp contrast to that carried on in the routine of official business. Born from the heat of disputation, this "rhetorical scholarship" was largely the creation of the Greek natural philosophers and Sophists and the pre-Ch'in (3rd century B.C.) philosophers of China.

People often speak of the "miracle of Greece." They marvel that Thales (624–546 B.C.) and the other pre-Socratic natural philosophers should have approached nature with a sensibility so radically different from that of the astrologers and court historians who dominated cultural life before them. It is less often observed that the sequence of their discussions about a primary "stuff" that underlies and produces all things is finally intelligible only as the product of controversy, only as the issue of impassioned debate. The abstract logic of the Eleatics and the atomic theory of Democritus (460–370 B.C.) had their inception neither in official documents nor in the incantations of diviners, but in the midst of disputation and in the course of relentless pursuit of the issues. It was in this context that recognizable differences emerged and individual thinkers came to be identified with particular theories.

There is a conventional view of the scientist as one who does his thinking alone in the quiet of his room, eschewing the polemics of the political arena. In ancient Greece, the nature of the cosmos was discussed with the same combative spirit as were political matters. Forged and honed in the direct democracy of the Greek city-state where citizens hammered out public policy through discussion and debate, this was a spirit that regarded words as sovereign and saw in persuasive techniques a means to personal advancement.

The natural philosophers themselves seem to have been essentially publicists, itinerant moralists who had been attracted to Athens from the many outlying Greek colonies by the city's wealth. In fact, there is some indication that it was only after their political ambitions were thwarted and as they faced persecution and banishment that they escaped to mathematics or disguised themselves as students of nature (Africa, 1964). In the eyes of the upstanding citizens of Athens, the natural philosophers remained outsiders who lived off the city.

Yet it is the Sophists who were regarded as social parasites by later generations. When Nishi Amane first translated "Sophist" into Japanese in the latter half of the 19th century, his study of the history of Western thought led him to choose the term *nise-gakusha* (僞学者) or "purveyors of false learning."[2] Today, however, a century of scholarship has done much to redeem the Sophists' reputation, and scholars are more willing to view them as serious—if condescending—educators (Hunt, 1961). When they are taken seriously and seen in the context of what is apt to be a conserva-

[2] Nishi Amane, *Hyakugaku Renkan*. Although this work was not published until much later, Nishi wrote the manuscript in 1870 and used it as the basis for lectures at his private school.

tive profession, their manner and methods appear quite striking, even revolutionary.

The stage and setting for the men who have been called history's first professional teachers was the Greek *polis* of the 5th century B.C.[3] Greek society was then at the height of its prosperity, and Athenian forms of democracy were spreading throughout the region, but public educational institutions were as yet unknown, and learning remained something that each individual had to arrange for himself. Drawn to Athens by its wealth, like the natural philosophers, the Sophists were to find their opportunity here. Not being citizens, they sought private patrons, advertising their knowledge in the marketplace and going from street to street in search of an audience and students. To these efforts the mild Athenian climate lent a helping hand, drawing people out of doors and fostering a fondness for conversation and discussion. Moreover, under a democratic regime, people with political ambitions were glad to pay the Sophists for their expertise in rhetoric. For their part, the Sophists delighted the crowds with their unusual doctrines and shrewdly cultivated their public image. Protagoras (5th century B.C.) is perhaps the best case in point, for he is said to have amassed a considerable fortune from the teaching fees he collected. Increased competition after his time led to a decline in the going rate for instruction and "initial" fees, and later generations did not prosper to the same degree. Nevertheless, Plato, Aristotle, Isocrates, and other "teachers of mankind" supported themselves by giving oral instruction and operating private academies. The education in which the Sophists were engaged was thus an individualistic activity in which teachers peddled their skill and would-be students paid for it. This was the atmosphere in which Greek learning was born. As a result, polemical techniques and the dialectical style came to be closely associated with Greek philosophy, science, and culture.

The persuasive and oratorical arts the Sophists left behind are of more interest than what they had to say. What matter a few errors, so long as they were able to get the better of their rivals! In the *Gorgias*, Plato's Socrates forces the orator to acknowledge

[3] In discussing the Sophists, I have generally adopted the view of the outstanding French historian H. I. Marrou. His views are available in English in *A History of Education in Antiquity*, trans. George Lamb (London, 1956).

that his rhetoric is not an art which teaches knowledge of right and wrong but merely a means of persuasion that engenders belief without knowledge. An art designed solely to convince, he argues, is a "spurious counterfeit." Rhetoric is not bad in itself, but it must serve the search for truth. These views of Plato sound quite similar to our own. But what surprises us as we read the *Gorgias* today is how absolutely indispensable oratory appears to have been to scholarship of the time. Socrates himself emerges as an exceptionally capable rhetorician. Moreover, rhetoric is regarded, at least implicitly, as a method of discovery. Polemical discourse stimulates the mind, activating and nurturing creative thought. The locus of true education is dialogue, for it is here that new ideas and patterns of thought emerge. At least this is what Socrates and Plato believed. Hence they preferred to entrust their thought to the living minds of men rather than to the skins of dead sheep.

The breakdown of the Greek *polis* and the establishment of the Macedonian Empire resulted in the geographical diffusion of what had been a regional Athenian culture. This posed a new challenge for Greek learning. Persuading people with different languages, customs, and manners required more than clever plays on words and the skillful use of oratorical techniques. Attention to grammar became necessary, and logic replaced rhetoric as the primary tool of persuasion. Substance also received increasing attention, for stripped of the power of rhetoric, vacuous remarks were more readily exposed as such. In Isocrates (436–338 B.C.) one already sees a man more interested in substantive matters than the rhetorical forms of the Sophists, and in Plato (427-347) and Aristotle (348–322) there is a definite shift of emphasis from mere rhetoric to logic.

The expanded context in which Athenian culture now sought to make its way also served to accentuate another element of the Greek argumentative style. If one is to be successful as roving orator, one's remarks must have sufficient generality or universality about them that they can be readily understood by anyone. Within the narrow confines of a sect, knowledge to be transmitted may be treated as authoritative in itself or as the "divine wisdom" of the founder. But the relationship between a teacher and an audience of strangers is more like that between a salesman and a group of skeptical customers. Neither sweeping claims to authority nor smug complacency are effective in such a situation—nor in any context involving diverse assemblies of people. Knowledge and ways of thinking that are developed in and shaped by discussions among diverse participants tend inevitably to be refined in the direction of the general and the universal.

Disputation begins as ordinary human conversation, with talk of things that can be seen with the eye and touched with the hand—with concrete things. As the discussion proceeds, however, it departs from concrete objects and events. Thoughts coalesce, convictions become clarified, points of agreement and conflict are articulated and urged, and the conversation grows more and more abstract. Thus abstraction is born of persuasive discourse. The fact that the ancient atomic theorists seem to have approximated the atomic theory of modern science without the aid of experimentation or experimental facts, even though they approached the problem with a different set of concerns, is an example of the results obtainable on occasion through the vigorous pursuit of abstraction in a polemical context.

Ever since the emergence of Hellenist civilization, there have been two streams of thought in the Western academic tradition: the philosophical school that began with Plato, and the rhetorical tradition founded by Isocrates as he drew upon and absorbed the work of the Sophists.[4] The latter stream has seldom received the attention accorded the former. Presumably less readily committed to writing, rhetoric suffered historically from the fact that it was not in keeping with the cultural tastes of early modern Europe. Long subjected to the ruthless attacks of the Platonists, the Sophists in particular emerged as the enemy of the "philosophers," retaining a place in the history of philosophy only as "bad guys." Yet the fact that they were better remunerated than either the Platonic geometers or the teachers of the "three R's" attests to the esteem they commanded in their society (Kōno, 1958). Nor can their lasting contributions be easily dismissed. They were

[4] We are accustomed to thinking of philosophers and rhetoricians as distinct groups, but in terms of their scholarly fields, no clear distinction can be made between the two. Mutual name-calling makes the distinction even more problematic. For instance, according to the "sophist" Isocrates, the Platonists were "sophists," while he and his followers were "philosophers." Thus one comes closer to the mark by referring to each of these scholarly groups by the name of the individual whose work was central to its tradition.

the precursors of grammar and linguistics (Forbes, 1933), and the forensic techniques that they left behind played a significant role in the formation of the Roman legal tradition (Periphanakis, 1953).[5]

During the same 5th century in which the Sophists prospered in Greece, the so-called "hundred philosophers" were actively propagating their views among the "Warring States" of China. The decline of the Chou Dynasty had given rise to a plethora of small, independent, contending kingdoms, and, in the midst of a political crisis not unlike that which occurred in ancient Greece, similar intellectual conditions also emerged. A lower ruling stratum known as *shih* (士) began to come into its own as the lords and nobles of these principalities sought to enhance their power and prestige by attracting the most able and accomplished men to their kingdoms. In response to this demand, powerful *shih* became lecturers to kings or high officials, and groups of lesser *shih* began to form around them. The *shih* were wandering scholars who sought to change the world with their words—and win appointment to office by selling rulers on the value of what they had to offer.

The amount of information these scholars could bring with them on their travels was perforce minimal. Astrologer-officials might preserve huge stacks of records, classified and indexed for ready accessibility, but the wandering scholar was limited to what he could carry in a straw basket strapped to his back. Before the invention of paper, the Chinese kept their records on foot-long strips of bamboo or wood. For all practical purposes this meant that the only records available to the wandering scholar were those he kept in his head.

Clearly the knowledge the wandering *shih* had for sale was not bound in quantities of documents. They were purveyors of ideas, not encyclopedia salesmen. If a prospective buyer happened to be intelligent, the encounter could also be intellectually stimulating. But the *shih*'s wares were not limited to ideas. Merchandising methods—rhetoric and logic—were marketed as well, as Legalists, Confucianists, and Mohists vied with one another to make their product as attractive as possible.

Inasmuch as they lived by persuasion, the interest in logic that

[5] This was first demonstrated by the famous 19th-century legal scholar F. K. Savigny.

emerged among the pre-Ch'in philosophers is hardly surprising. The conceptual analysis of Kung-sun Lung in his famous *Discourse on the White Horse* and the paradoxes of Hui Shin that appear in the *Chuang-tzu* are the work of an early group of logicians (often known collectively as the *Ming-chia*) and fully worthy of the term "sophist." A concern with logic is also evident in Mohist writings, as are the seeds of development in additional areas otherwise rare in the Chinese tradition—mechanics, geometry, optics, and epistemology.[6] In Chuang-tzu, the period also produced a thinker with an expansive, almost breathtaking view of the cosmos.

A more careful look at the Chinese rhetorical tradition, however, discloses a preference for arguing in terms of precedent and previous example that contrasts with the Greek style of logical persuasion. This tendency was reinforced in the Han period by Emperor Han Wu Ti's exclusive preference for Confucian teachings. Under his regime, the teachings of all other schools were suppressed. Even Mohism, which once had divided the intellectual world with the followers of Confucius, all but died out, not to be revived until the early 20th century when intellectuals under the influence of modern Western science welcomed its radical critique of Confucianism.[7]

Propensities and Functions

The distinctive character of the two scholarly styles should now be clear. The documentation of extraordinary occurrences in the heavens is cast in a settled, agricultural mode. The work is steady and the yield is stable, but the product is plain, restricted in scope, and conservatively inclined. The rhetorical style of the Sophists and pre-Ch'in debating scholars, on the other hand, is of the mobile, commercial type. It is pursued with a broader vision and tends to be progressive, but it cannot exist without an adversary and leads to the accumulation of substantive knowledge only in exceptional cases. New theories appear one after another, but rise and fall in such a manner that accumulation can be little more than another name for confusion.

[6] See the book of *Mo-tzu*, the "Ching P'ien" and the "Chang Shuo P'ien" sections.
[7] The best known work of the Mohist revival is by Hu Shih (1928).

Scholarship in the documentary style lacks the capacity to generate new problems. And when new problems are not forthcoming, the enthusiasm of scholars declines and their work goes flat. Rhetorical scholarship, on the other hand, is concerned with nothing if not the creation of problems. Posing and counterposing questions is its major function. But disputation seldom leads to firm conclusions. The adversaries advance their own doctrines and strive to persuade opponents of their own veracity. Such encounters provide opportunities for participants to refine their opinions, but the attainment of a "truth beyond" that transcends the particular views of the participants involved is not its object. Thus, whatever may eventually emerge from such discourse, it does not become generally accepted "truth." Indeed, the outcome is normally inconclusive. When interest wanes and the participants leave the scene, the controversy vanishes into thin air without becoming an academic tradition.

Documentary scholarship has as its proper object certain objective phenomena it is obliged to record, but scholarship in the rhetorical style is not likewise bound, though the polemicist may be constrained by his opponent. Thus its discussions occasionally digress from the subject at hand and end up as abstract debates over empty theories. Those who prefer proof to precept will always find such discussions unsatisfactory. And yet abstraction has always been a proper function of argumentative discourse.

During the early period with which I have been concerned in this chapter, these two styles of scholarship were pursued largely without reference to each other. Disputation did not necessarily proceed on the basis of previously recorded data, as Plato's famous distinction between knowledge and opinion reminds us. Recent scholarship has increasingly embraced the view that the "miracle of Greece" was really no miracle at all, that Greek natural science was in fact largely built upon data accumulated by, and borrowed from, the empirical science of Babylon (Neugebauer, 1953). Yet it was a phenomenon of the late classical and early Hellenist period (cf. Chapter 2). The rhetorical scholarship of the Greek natural philosophers and Sophists neither required nor presupposed the Babylonian records of extraordinary occurrences in the heavens, and Greek influence on these records is inconceivable.

A mass of detailed, particular facts may supply the polemicist with data for argument, but one may point out many instances in the history of science and learning in which the weight of accumulated documents had become so great as to inhibit the emergence of a theory able to control all available data. An abstract notion such as the idea that all things come of water cannot be formulated while the mind is wholly absorbed in the details of things and events. Conversely, Chinese government records are said to have been rife with the controversies of the hundred philosophers before being edited as official histories.

Which of these two styles of scholarship should be called learning? Which deserves the name science? Although I have thus far seemed to view the rhetorical style slightly more sympathetically, this question cannot be answered unequivocally. It is, of course, easy enough to say that learning encompasses both functions and that healthy growth occurs when each type functions fully and a balance is maintained between them. But how is this balance to be maintained? And what is healthy growth? From the lofty heights of modern science, a kind of "victor's logic" readily leads us to assume that the course of Western science precisely reflects this balanced and healthy development. Yet such a view has no sanction. Moreover, insofar as our task is not to celebrate modern science but to understand from a comparative perspective the dynamic historical changes that have made the world of learning what it is today, such preconceptions can only impede. It will be more fruitful to ask not which scholarly style is better but rather how each has functioned in the history of science and learning wherever we find it.

Documentary scholarship finds little meaning in repeated recording of the same phenomenon; it is variety that is welcomed. Extraordinary phenomena are particularly worth documenting. It is such phenomena rather than the discovery of laws that constitute the chief concern of the astrologer and the historian. The ancients regarded both lunar and solar eclipses as extraordinary occurrences and recorded them. Since the prediction of lunar eclipses turned out to be comparatively simple, their nomological character was soon discovered. Once this occurred, astrologers lost interest and ceased to record them with any regularity. Solar eclipses, however, proved more difficult to forecast, and they continued to attract the attention of observers.

Even among the rhetorical traditions one can recognize degrees in the relative importance attached to documentation and demonstration. The contrast between the dialectic of the Greek philosophers and the tendency of the ancient Chinese philosophical schools to employ historical example as a persuasive device has already been noted (Graham, 1973). But differences obviously existed among the Greeks as well. Though extant records might well suggest otherwise, the first to advocate the preservation of knowledge in books was not Plato, whose quest for universal truth was accompanied by a distaste for writing, but the Sophists, with their interest in particular phenomena and their love of rhetoric.

As they became embroiled in actual controversies, the Sophists and the followers of Isocrates did of course strive to cast their remarks in general, universal terms. Yet they were clearly less thorough than the Platonists in this respect. Here one can already see two distinctive intellectual temperaments at work. The Sophists held that the world is in a state of constant change and flux. As relativists, they were interested in change, in variety, and in the particular. Avoiding the strict adherence to things eternal that dulls sensitivity to the actual, they cultivated a direct, empirical, pragmatic, and opportunistic attitude toward the changing world they saw around them. And compared to those who sought coherence, universality, and eternity, to those whose quest for law and order was liable to make them conservative, the Sophists were progressive, even radical.

The quarrel between sophist rhetoricians and philosophical polemicists has appeared in many guises throughout the course of history (Streuver, 1970), and still continues today. Historians, publicists, and journalists tend to be arrayed on the change-oriented side. This type does not seek eternal, unchanging laws, for it knows them to describe only the prosaic, the commonplace, the recurring. It responds enthusiastically only to the exceptional or extraordinary. Those who pursue unchanging regularities, on the other hand, are driven by a theorizing, systematizing impulse. They would reduce the actual diversity of the world to a series of generalizations, stop the flow of history at a given point in time, and force the world into a mold of their own making. As disturbed by the exceptional as the change-oriented are delighted, they are as a type inevitably oriented toward order.

The Sophists gave birth to the argumentative arts of grammar, rhetoric, and logic (the so-called Trivium), and cultivated a style that sometimes reminds us of contemporary social scientists. But one rarely finds among them an interest in natural science; indeed, they showed little potential for development in this direction. When the rise of Hellenistic despotism deprived them of the political forum they had enjoyed during the classical period, the heirs to the Sophist-rhetorical style turned to narrative history and became the forerunners of classical studies in the West (Pfeiffer, 1968). Meanwhile the Platonists moved toward a greater concern with demonstration and evidence and developed new interests in natural philosophy and natural science. But this is a subject for the next chapter.

Chapter 2

The Emergence of Paradigms

The word "paradigm" first came to the attention of students of science and civilization with the publication of Thomas Kuhn's widely read and much-discussed study, *The Structure of Scientific Revolutions* (1970). Making the concept central to his analysis of scientific development, he employed it to designate "generally accepted scientific achievements that, for a time, supply some particular scientific community with a model for the formulation and resolution of research problems." Newton's *Principia* (*Philosophiae Naturalis Principia Mathematica*) is perhaps the best known and most typical example of what Kuhn meant by a paradigm. As he observed, this work not only served as a model for subsequent research in the field of mechanics, but also exerted a pervasive influence throughout the natural sciences, as the mechanistic methods and mechanistic conception of nature exemplified in its pages spread to chemistry, biology, and other fields.

Kuhn's work was first published in 1962, and his portrait of science and scientific revolutions soon became the subject of considerable controversy.[1] Some of the difficulties raised by critics can be traced to the fact that Kuhn used the paradigm concept in both a methodological and a sociological sense without making this duality fully explicit. But some philosophers of science also assailed the ambiguity of its applications to the philosophical and theoretical dimensions of science, and Kuhn has now felt obliged to withdraw it in favor of more restricted terminology. Yet in this concept the author has given the scholarly world a much-sought-

[1] Kuhn's paradigm concept has been discussed in countless articles. One attempt to pull some of this discussion together is David H. Hollinger (1973). M. D. King (1971) is worth reading as one of the early discussions of Kuhn's work by a sociologist. These are all reprinted in Gutting (1980).

after tool of reflection and analysis.[2] It is now so widely used that it has a life of its own. In a sense that I will shortly define, the term will be used in this work to identify a constitutive principle of scientific groups, sociologically and historically considered.

There are a variety of specialized, professional groups in existence today. These groups vary in size and scope and often appear to be constantly proliferating by fission. Thus one may speak of natural scientists, physicists, nuclear researchers, etc. In identifying these groups as groups, we imply that there is something distinctive about each of them—something to which each owes its origins as a historical community, its development and elaboration as a mature science, and its continuity as a social entity. This "something" forms a kind of "elemental bond," a tie shared by all members of the group. It is this basic characteristic of specialist groups that constitutes the first precondition for the establishment of paradigms.

The members of such groups often have many things in common: race, nationality, language, basic articles of faith, education, and so on. That there are many different kinds of social groups is equally obvious. What then is the nature of the "elemental bond" peculiar to groups of scholars and researchers? Here the notion of a generally accepted basic achievement alone is insufficient, for religious groups have something comparable in their scriptures, and programs, platforms, and slogans perform a similar function in political parties. If the term "paradigm" is to be used solely with reference to the kind of bond peculiar to scholarly groups, it must designate not merely a basic achievement but one that, more specifically, *provides its supporters and their successors with*

[2] See Kuhn's "Postscript—1969" in the 1970 edition of *The Structure of Scientific Revolutions*. Imre Lakatos and A. Musgrave (1970) contains some of the best criticism of Kuhn's position. Opponents have generally made two main points. The first is that the term "paradigm" gives rise to misunderstanding because it is burdened with everyday meanings such as "example." The other is its multiple meanings. The first objection can be easily taken care of simply by creating a completely new word. All that would be required of the term is that it not be reducible to existing terminology such as "theory" or "hypothesis," for "paradigm" actually seeks to express a new understanding of scientific activity. If research is actually carried on less as an exercise in the conscious application of some theory of science or the scientific method than through picturing in the mind some sort of concrete, previously completed, exemplary model, then one comes closer to the sensibility of the researcher by considering paradigms as undivided wholes (i.e., a particular book, research example, or instrument) rather than as possessing a logical consistency that would satisfy the philosophical analyst.

a point of departure that assures the development and resolution of the several types of problems with which they are confronted.

Defined in this way, the notion of a paradigm can be applied to the study of fields beyond the natural sciences. One thinks immediately of the role which the writings of Karl Marx have played for a large group of scholars in the social sciences. Grand views of history perform a similar paradigmatic function in historical studies. Nor should a list of paradigm candidates be limited to achievements that have taken the form of theories or books. The telescope, the microscope, and other revolutionary instruments such as the computer can be called paradigms because they have opened up new fields of inquiry. The same might be said of mathematical methods such as the analytical calculus. Basic technological inventions such as the steam and jet engines also deserve to be called paradigms inasmuch as they have led to the formation of research groups engaged in improving the performance or expanding the uses of the inventions.

The above are major paradigms of the sort that mark a new beginning in the history of science and technology. Yet consideration of even the most specialized fields should yield some kind of minor paradigm to which each owes its identity as an academic tradition. The term can also be applied more broadly to achievements that transcend particular fields, achievements that operate at a different level to define a particular academic style and provide the fundamental modes and systems by which knowledge comes to be ordered and arranged; two examples are the works of Aristotle and the Confucian classics. In this case paradigm signifies *canonical codifications of classical texts that set scholarly style, legitimate the specialized, professional activities of intellectual groups, and lay out a course of subsequent development for what has come to be normatively defined as scholarly activity*. In this comparatives study of academic traditions, I shall be using the term in this larger sense.

If paradigms may be thought of in this way, they will be seen as performing a function that in the philosophy of science is currently reserved for the abstractions it terms the first principles or rules of scientific method. The reason I have used the term "paradigm" to designate this function is that academic research actually proceeds in a much more flexible manner than is suggested by such terms as "principles" or "rules." The researcher normally models his work on a specific piece of previous research, proceeding more by trial and error than by the strict, mechanical application of abstract scientific laws. Through the study of Marx's writings, the Marxist is less interested in understanding theories and concepts such as socialism, materialism, or the dialectic than in learning a way of thinking, of raising questions, and of framing answers.

Normal Science

Before a particular mode of learning acquires settled procedures and subject matter, inquiry takes place amidst an anarchic clash of theories reminiscent of the rhetorical style described in the previous chapter. The confusing array of explanations of cosmic motion that were available prior to the publication of Newton's *Principia* is an excellent illustration of this "state of anarchy." Keplerian dynamics and the Cartesian theory of vortexes mingled with traditional Aristotelian teleogical views and Galileo's kinetic explanation. Such conditions offer little hope for normal "progress" and "development," for a scholar who does not know which theory to follow in his research cannot contribute to the advancement of knowledge. Yet to advance a new theory of one's own at this stage is merely to add to the confusion; in itself, it can by no means be called progress.

But anarchic confusion cannot be endured forever. In time one theory or method wins out over others and gains acceptance among researchers as "the standard theory." To acquire this status, an opinion or practice must appeal sufficiently to a group of supporters that they abandon other ways of working and embrace it as their paradigm. Yet once established and accepted, it occupies an authoritative position in the minds of succeeding generations of scholars, providing them with a set of standard formulas and basic research tactics. It is no doubt destined to be replaced someday, but for the time being acceptance of the paradigm means that scholars can proceed with their research in the confidence that they are operating from a reliable base.

The establishment of a paradigm is followed by a wave of research modeled upon it. Since it is no longer necessary to rehash earlier controversies over basic questions, revolutionary discoveries are succeeded by a steady accumulation of work along the lines suggested by and set forth in the paradigm. The result is "normal science," a chain of ongoing scientific research patterned after an exemplary paradigm and built upon its authority.

A hypothetical example will illustrate how the establishment of a paradigm marks the transition from disputation to normal science. Let us suppose that while working with hydrogen someone discovered a radically new method of studying the elements. This method would soon be applied paradigmatically to helium and the other hundred or so elements in the periodic table, generating similar problems and producing a succession of significant results. A tradition of normal science has now come into being, though such traditions may be much broader in scope than is suggested by this illustration.

Conducted with the assurance that work properly done will yield a satisfactory answer, the work of normal science can hardly be described as a romantic "quest for the unknown" or an "uncharted intellectual adventure." On the contrary, the search in which up to ninety-nine percent of all scientists are actually engaged is a search for the patently knowable.

The documentary learning discussed in Chapter 1 may be seen as the earliest form of normal research. If this scholarship can be said to have a paradigm it is the recording of precedents. In the case of astrological records of extraordinary celestial phenomena, the paradigm dictates the kinds of phenomena that are considered extraordinary, the style in which they should be recorded, and the way the collected records are to be classified. Baconesque natural histories are also a form of normal science. And normal science by definition makes cumulative "progress."

Quantitatively, one can gauge the scope of a given paradigm by the number of normal science problems it conceives and the number of scientists engaged in their resolution. Yet, whatever the scope, when these problems have been resolved and its capacity to generate new ones has been exhausted, the paradigm ceases to function. When this happens the scholarly tradition founded upon it dies, and the information it has produced is fossilized and preserved in textbooks and encyclopedias.

Scientific Revolutions

When an established tradition of normal science encounters phenomena it is incapable of explaining, the tradition is plunged into crisis. Doubts about its foundations are reborn and controversies arise. Eventually a different paradigm is adapted and the style of scholarly activity undergoes a transformation. When this happens on a major scale we say that a scientific revolution has occurred. Both the triumph of quantum theory over classical physics and the replacement of Newtonian mechanics by the theory of relativity may be seen as scientific revolutions that came about through the emergence of new paradigms.

Though employed in a variety of contexts, the word "revolution" is preeminently a sociopolitical term. With the emergence of new elements and new classes, contradictions appear in the old order. If the established forces remain strong, the old order can be partially renovated and the several contradictions absorbed within it. But when contradictions exceeding the capabilities of piecemeal revision come to the fore, the need is felt for a new order. During this time of searching, a state of anarchy sometimes arises, but after a period of struggle and controversy among groups advocating a variety of sociopolitical models, a new order begins to take root and people make their peace with it.

In speaking of the transformations that accompany paradigm changes as scientific revolutions, I intend to suggest that the scholarly paradigm be regarded as functionally analogous in certain important respects to a sociopolitical order or educational system. Thus when the old paradigm ceases to function effectively, the groups that are committed to it collapse and a revolution occurs. With the fall of the paradigm, the basic articles of faith it celebrates, the language (conceptual system) it employs, and the teaching methods it uses are swept away and replaced by a different set.

The shift from Ptolemaic geocentric astronomy to the heliocentric astronomy of Copernicus is one of the best examples of this kind of major scientific revolution. The paradigm of earth-centered astronomy was Ptolemy's *Almagest,* a comprehensive survey of Greek astronomy composed in the second century A.D. This work expressed all the movements of the heavenly bodies as a series of uniform circular motions. Where these cycles were not sufficient to account for observed phenomena, they were supplemented by epicycles (or in some cases eccentric circles). Proper adjustments of the radii and rotational velocity of these epicycles made possible the expression of almost all movements. In theory, epicycles could be added *ad infinitum* until every phenomenon had been accommodated. This was in fact precisely the direction Ptolemaic astronomy took as it developed in a normal fashion; and in this direction, cumulative progress was certainly made.

Why then did Copernicus find it necessary to embark on a revolutionary course? To be sure, contemporary astronomical calculations were in need of revision. Periodic revisions are regularly required when using older calculations, for some elements inevitably grow increasingly inaccurate with the passage of time. Even the most precise clock must sooner or later be reset if it is to be an effective timepiece. If accuracy had been the only issue, modifications within the framework of the Ptolemaic paradigm would have sufficed.

Copernicus, however, was troubled by a more profound matter. With a proper combination of epicycles one could indeed "save" any phenomenon. Moreover, to this end the tradition permitted considerable latitude in the way in which the number of epicycles, their radii, and the velocity of their revolutions could be combined. But precisely because all that mattered was that the scheme accord with observed phenomena, there were no conclusive criteria for determining whose method was correct. So many approaches were advocated that, in time, a state of intellectual anarchy existed among those committed to the same Ptolemaic paradigm. Under such circumstances, the *theoretical* value of the Ptolemaic "theory" was questioned. In the Preface and Dedication to Pope Paul III affixed to his *De Revolutionibus Orbium Coelistrium* (1543), Copernicus makes the following observations about the Ptolemaic "mathematicians":

> Even if those who have conceived eccentric circles seem to have been able for the most part to compute the apparent movements numerically by those means, they have in the meanwhile admitted a great deal which seems to contradict the first principles of regularity of movement. Moreover they

> have not been able to discover or to infer the chief point of all, namely, the form of the world and the certain commensurability of its parts. But they are in exactly the same fix as someone taking from different places hands, feet, head, and the other limbs—shaped very beautifully but not with reference to one body and without correspondence to another—so that such parts made up a monster rather than a man.[3]

He went on to observe:

> Accordingly, when I meditated upon this lack of certitude in the traditional mathematics concerning the composition of the movements of the spheres of the world, I began to be annoyed that the philosophers who in other respects had made a very careful scrutiny of the least details of the world, had discovered no sure scheme from the movements of the machinery of the world, which has been built for us by the Best and Most Orderly Workman of all.

In order to lay waste this monster, Copernicus devised a heliocentric theory which, among other things, enabled him to do away with certain epicycles. Kepler subsequently replaced the circle with an ellipse, making uniform motion unequal and abolishing epicycles altogether. Judged in terms of its contribution to the cumulative growth of epicyclic astronomy, this development was a step backward. Yet from the standpoint of Copernican and Keplerian astronomy, the devoted labors of Ptolemaic astronomers to enhance and perfect the epicycle theory seem to have been leading nowhere—wasted effort, expended in the wrong direction.

Between the two there was a shift in the immediate and specific goals of scholarship that concerned itself with the "movements of the spheres of the world." A scientific revolution had taken place, and a new astronomy, paradigmatically grounded in Copernicus' heliocentric theory, began its life as a normal science (cf. Figure 2–1). A tradition of normal science radically different from

[3] Nicolaus Copernicus, *On the Revolutions of the Heavenly Spheres*. The passages quoted here are taken from the English translation of Charles Glenn Willis, *Great Books of the Western World* 16 (Chicago, 1952).

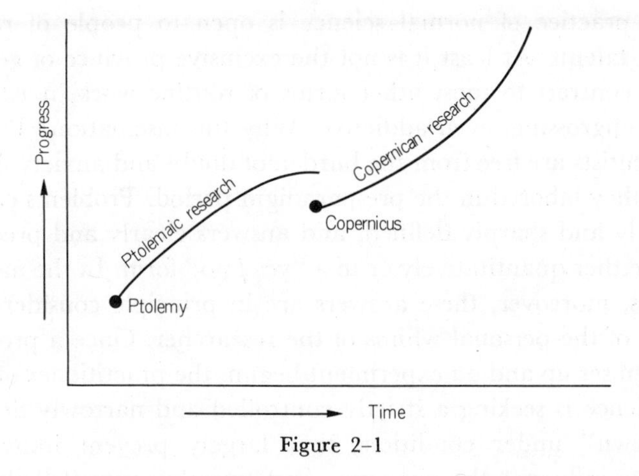

Figure 2–1

those which had preceded it had emerged from and through a paradigm revolution. The cycle was now complete: (old) normal science—contradictions—paradigm crisis—scientific revolution—adoption of a new paradigm—new normal science.

Revolutions and Normal Science

The revolutionary work of creating a new paradigm and the accomplishments of ordinary science are, of course, both scholarly achievements, but there is a distinct difference between them that should not be forgotten. The great bulk of contemporary research in the natural sciences is conducted along the lines laid down by a particular tradition of normal science. Those engaged in this kind of research do not ordinarily raise doubts about the fundamental principles of their science. Indeed, the distinguishing feature of normal science is that it proceeds along a pre-established course, constantly accumulating knowledge to be added to what has already been confirmed. Once a paradigm has been selected and an orientation set, the track is laid and the research group is ready to speed off into the distance. It is this swift, seemingly unproblematic, and apparently endless character of its advance that sets modern science apart from other modes of learning and often arouses their envy. This cumulative-style progress has prompted the social sciences and other disciplines to model themselves on modern science, particularly modern physics.

The practice of normal science is open to people of rather modest talents. At least it is not the exclusive province of genius. Yet in contrast to most other forms of routine work, it is often deeply engrossing, even addictive. Why this fascination? First of all, scientists are free from the burden of doubt and anxiety under which they labored in the pre-paradigm period. Problems can be narrowly and sharply defined, and answers clearly and precisely given, either quantitatively or in a "yes / no" form. In the natural sciences, moreover, these answers are in principle considered to be free of the personal whims of the researcher. Once a problem has been set up and an experiment begun, the practitioner of normal science is seeking a strictly controlled and narrowly defined "unknown" under conditions that largely prevent individual biases to influence the outcome. And, strictly controlled though it is, the discovery of this unknown promises a thrill not unlike that of solving a puzzle, a prospect so consuming that the researcher may forget to eat until the experiment has been completed and initial calculations have been made. This capacity to provide an abundance of puzzle-like, clearly delineated problems and a research apparatus by which they can be resolved has been one of the key factors behind the exponential growth of modern science.

Beyond these immediate gratifications, the pursuit of normal science offers even larger rewards. First, the researcher makes a contribution to the storehouse of human knowledge, work to which no one can object. Through participation in this cumulative enterprise, the individual can feel that his life is being well spent. Finally, because the incremental progress represented by any given achievement is measurable, normal science provides a framework within which the researcher can acquire the social respect accorded to those of intellectual achievement.

But what does the researcher do when something unusual appears in the course of a normal research project set up in accordance with the prescriptions of a given paradigm—something abnormal that seems to run counter to the paradigm's predictions? How does he deal, for instance, with a situation in which he encounters observational facts that cannot be explained by Newtonian dynamics? A normal science practitioner's first thought will be that he has erred in his observations, and he will scrutinize his materials and methods carefully to make sure they are in keeping with established theory. But suppose that all such efforts fail to resolve the problem What then? Had he the makings of a revolutionary, he would make positive use of the abnormal phenomenon as a means of breaking through the impasse its appearance created. That is, he would formulate a new paradigm capable of dealing with the phenomenon and lay the foundations of a new tradition of normal science.

Once embarked on the task of overturning an established tradition, however, neither the researcher nor his friends and colleagues will be able to tell whether what he is doing is constructive or not. For at this point he no longer has the assurance that he is accumulating knowledge and thereby participating in scientific progress in the ordinary sense of those terms. His training in the scientific method is of little help, and sometimes actually works to check the revolutionary impulse. Moreover, inasmuch as the revolutionary would replace a paradigm that is the focus of commitment for an established group of normal scientists with one of his own making, he finds himself constantly at the center of controversy. Buffeted by criticism, he may be subject to frequent periods of mental crisis during which he becomes unsure of what he is doing and loses confidence in himself. At times, extra-intellectual resources are required to withstand the pressures lest, like the Austrian physicist Ludwig Boltzmann, he be driven to take his own life. Hence, the lives of paradigm makers are often tragic and lend themselves readily to heroic biography. The advancement of science may be the work of ordinary people, but it is geniuses who turn it around and change its direction.

Nevertheless, it is not creation but progress that is the distinctive feature of science. Only when the paradigm has evolved into a normal science program can one speak of science. Unless this is recognized—unless the history of science is seen as something more than the story of creative, paradigmatic achievements—it will not be possible to either grasp or portray the awesome power of normal science. In modern science, research can proceed without going back and scrutinizing Newton's *Principia*—and this is progress.

The Social History of a Paradigmatic Achievement

I have discussed the meaning of paradigmatic achievements and surveyed their crucial role both in the scientific revolutions that marked the rise of modern science and, more particularly, in the constitution of scientific groups and the practice of normal science. Now I will attempt to relate these observations to the external factors that have been involved in the formation of scholarly groups, to the sociological aspects in the history of a paradigm.[4] Let me first present a general model.

Paradigms originate in the minds of individual men and women. It would be foolish to argue that the social environment has no role in this process, but the efforts of Marxist historians of science to show that society is the decisive determinant of thought (that, for instance, Charles Darwin's theory of natural selection was simply a reflection of the unrestrained competitiveness of Victorian society) have not been notably successful (Bernal, 1954: Chs. 6, 9). The individual is always subject to the workings of his environment, but an analysis of the social factors involved and the way in which they function at the precise moment an idea is born would appear to be nearly impossible. Should the idea that the earth revolves around the sun emerge in a primitive society, however, it would soon perish for want of anyone to appreciate its significance and pursue it. In this sense, unconventional ideas and modes of thought most often make their appearance in transitional periods (in either national or academic societies) under favorable environmental conditions. But which of the many ideas expressed during these periods will survive and take root as a paradigm is determined at another stage by another generation—by the successors who make the selection.

Historically this process seems to resemble the evolutionary principle of natural selection. Amidst the fluctuating conditions that marked social life in classical Greece and Warring States China, a variety of different views, values, and methods found expression. During the more stable Ch'in-Han and Hellenist periods that followed, doctrines considered appropriate to the new

[4] I have deliberately used the word "sociological" here rather than "social" because I want to deal not with national societies but with small groups of researchers as societies in themselves, and hence as social phenomena.

setting won out in the struggle for existence and came to be adopted as paradigms. When the history of learning is viewed from this perspective, Aristotle and Confucius as individuals are less important than those who came to regard them as the bearers of charismatic gifts, as founding fathers.[5]

The significance of this observation looms even larger when one considers that, when confronted with the conflicting theories and doctrines of several would-be founding fathers, the choice of a paradigmatic figure can seldom be made solely by asking which one seems more reasonable or logical.

Whether a certain doctrine—a view of history, a conception of nature, a line of inquiry—is correct or not is a question that does not resolve itself overnight. The appeal of a given paradigmatic achievement, the extent to which it will yield attractive problems and lead to fruitful results, can be determined only in the course of subsequent history. Thus a satisfactory answer to the question of why Aristotelian learning came to occupy a central position in the West and why Confucianism took root in China cannot be found in the writings of Aristotle and Confucius alone. It must be sought rather among those who selected and endeavored to perpetuate their scholarly traditions. And this is a sociological question—a question of the character and concerns of the advocate group. Even if it can be maintained that ideas are born spontaneously, time and place exercise a decisive influence on the shape of the advocate group and the mechanism of the selection process.

The advocate groups of which I have been speaking are, in the first instance, groups of *disciples* (a term that is still operative today in the word "discipline")—the academic associations, research groups and similar bodies concerned with the development of a new field or mode of learning. (These groups will be discussed in Chapter 4.) When learning has become formalized enough to be regarded as a discipline, however, a new stage has been reached. The initial disciples have direct contact with the master, but with the passage of time this becomes impossible. In order to facilitate the transmission of the paradigmatic doctrines to which they are now committed, succeeding generations find it necessary to codify

[5] It has been suggested that to treat history as a process of choice rather than as a source of origins enhances its contemporary significance. See Lynd (1968).

these teachings in a way that emphasizes certain aspects and slights others. This molding of the paradigm into a more readily assimilable form, this tailoring for educational purposes, is one stage at which social factors make themselves felt.

The common desire of the advocate group to perpetuate itself is a process of professionalization that also invites the influence of external factors. In order to make a living with a particular academic discipline, professionals make all kinds of compromises with and adjustments to existing social requirements—compromises that affect the shape and character of the learning they profess. For it is only when the advocate group with its codified mode of learning succeeds in acquiring a stable means of replenishing itself that the social history of a paradigmatic achievement is complete and the tradition it spawned becomes a recognized semi-permanent social force. And even here, in the professional groups and educational institutions (to be distinguished from freer academic societies and research institutions) in which the tradition has now found a home, all kinds of extra-academic local interests are at work, interests that affect the nature of scholarship and of scholarly activity itself. In the modern period, the university (to be treated in Chapter 5) is the primary locus of the process.

The course traversed by a paradigmatic achievement as it moves from the individual mind where it was conceived to institutional expression or accommodation may be graphically depicted as in Figure 2–2.

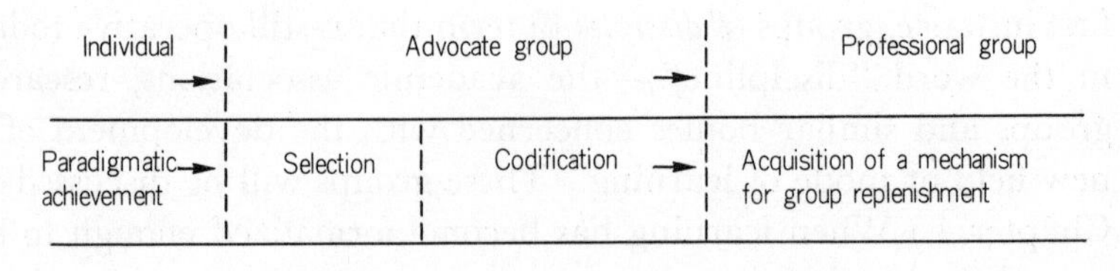

Figure 2–2

This scheme is not advanced as the only process by which paradigms can or have come into being as social realities in the course of history. It is merely an idealized pattern incorporating the major historical moments in the life of a paradigm in order to facilitate comparative discussion of cases. In some of these cases the order is actually reversed, educational institutions being established first and paradigms being brought in from outside and transplanted within them. This inverted pattern can be seen both in 19th-century Japan and in the modernization of underdeveloped countries in more recent decades. I will take up this case in Chapter 6.

The Historical Foundations of Scholarship: A Comparative Overview

When inquiring into the origins of a particular concept in the history of Western thought, one first goes back at least as far as Aristotle to see what he might have had to say about the subject. In the Chinese cultural sphere, one asks whether the idea in question can be found in the Four Books and Five Classics of the Confucian canon. In this routine practice, scholars acknowledge that the basic tone of the Western academic tradition was set by Aristotle while that of the East had its origins in the teachings of Confucius. The achievements associated with these two men may thus be regarded as paradigms of the highest order.

Of course fundamental ideas cannot really be counted, but some scholars estimate that two-thirds of the scholarly paradigms in use today originated in classical Greece. The question we must ask, then, is why, among the multitude of theories and arguments afloat in classical Greece and Warring States China, those of Aristotle and Confucius assumed the paradigmatic mantle. To be sure, both were exceptionally gifted individuals. But they were hardly the only superior minds of their times. In terms of the power to attract an organized band of followers, both Pythagoras in the West and Mo-tzu in the East also seem to have done quite well, and at the height of the ancient controversies and debates, no one could have predicted the final victor.

Interestingly enough, paradigmatic achievements were codified, providing canonic texts in many fields of academic endeavor and setting the larger patterns of scholarship, at approximately the same time in both East and West. While the Hellenist world (4th century B.C.—1st century A.D.) was compiling, framing, and ordering earlier scholarship within the framework of the Aristotelian tradition, the Confucian texts were gaining recognition as the authoritative rule of scholarly faith and practice in Han China (3rd century B.C.—3rd century A.D.). In both cases this development took place as stabler, more autocratic regimes came into being. It was in this sociopolitical environment that oral disputation gave way to a concern with a written canon.

In the West this trend is visible in the two major achievements to which we are chiefly indebted for our knowledge of the ancient Greek thinkers: a collection of Plato's dialogues compiled by the first generation of his disciples and a compendium of Aristotle's writings put together under the auspices of the Peripatetic School. As already noted, the Sophists had previously emphasized the significance of writing down orally delivered theories and arguments (Pfeiffer, 1968: 98), but it was only with the works of Plato's and Aristotle's disciples that Greek learning really entered the age of books. Since these disciples were chiefly concerned with preserving the doctrines of their teachers, they inevitably viewed the teachings of other schools through Hellenist and even Aristotelian eyes. Whether they accurately represented these teachings is open to doubt. In the Chinese case, we certainly do not have a full account of the pre-Ch'in thinkers, for the Han Confucianists who compiled and edited the canonical tests willfully distorted what they considered heterodox doctrines. In other words, the sources for the study of rhetorical learning in the ancient world are Hellenist and Han versions.

Since the ages of Confucius or Socrates and Plato predate the invention of paper, preserving lengthy debates in writing was a laborious task. Nor were the ancient practitioners of rhetorical learning inclined to record their opinions in writing. Socrates's views are reported in the Platonic Dialogues, but Socrates himself did not write anything; nor did Plato attach much significance to writing. He has even been said to have possessed some lost esoteric wisdom (like that of the Pythagoreans) that he only communicated verbally to a few of his most intimate disciples (Wherli, 1960; Kramer, 1959). It is not until we come to Aristotle, then, that we find a deliberate attempt to preserve the glory of the Greek past by collecting old documents. In this sense Aristotle was both the last classical Greek and the first great archivist, the last man to convey the whole spirit of the ancient debates and the first whose approach to scholarship clearly reflected its passing (Pfeiffer, 1968:

67, 87). Plato's *Dialogues* are peopled with many of the colorful figures who participated in the Greek argumentative tradition, but Aristotle treats the tradition with a new detachment. In his writings he makes every effort to present a variety of diverse and opposing views—including those of Democritus and the Ionian natural philosophers whom Plato kept at a respectful distance—and establishes an academic style that strives for precision and clarity even when the author's own opinions are involved. Francis Bacon found him "fond of disputation," but it may well have been Bacon's comprehensive survey of Greek scholarship that saved rhetorical learning from oblivion and assured it a permanent and articulate place in the Western academic tradition.

One of the distinctive features of Aristotelian scholarship was its classification of knowledge, which made rigorous distinctions among the paradigms peculiar to each scholarly field. This systematic ordering of knowledge was to become the prototype of higher learning in the West, and in its major divisions—natural philosophy (physics), astronomy, biology, logic, metaphysics, ethics, and politics—it is still being followed today. That Aristotle was long considered the father of all learning (notably in medieval Islam and in the late European Middle Ages) may be attributed to the fact that he was the first person in history, as Bertrand Russell put it, to "write like a professor"—the first to have employed a systematically organized, encyclopedic academic style that went beyond fragmentary knowledge and disputation, aphorisms and dialogue (Russell, 1945). That is to say, the professorial style that became a major element in the definition of learning developed in the West was an important ingredient in the codification of Aristotle's learning by his advocate groups. When Theophrastus founded the doxographical tradition with a collection of the opinions of the classical natural philosophers from Thales to Aristotle and another disciple, Eudaemus, assembled other kinds of materials, they were contributing to the growing dignity of classical studies in the Hellenist intellectual world.

Seeking to explain the ascendancy of Aristotelian rhetoric, the lst-century writer Quintinius observed that, though there had formerly been two schools of oratory, the Isocratic and the Aristotelian, the latter had survived because the Peripatetics had reinforced it and given it a system. Yet in so doing they were also following their master, for Aristotle had made oratory philosophically respectable by grounding it in logic (Solmsen, 1941), and had created a demonstrative natural science by linking it to the study of nature.

When, after the political instability of the late classical period, order was restored early in the 3rd century B.C., the freedom of the city-state vanished. Political discourse declined and a new type of scholar was born—meek, short on originality, and obedient to the established order. Though occasionally chided for their docility—the Alexandrian scholars were sometimes referred to as "pet canaries" (Pfeiffer, 1968: 98)—these scholars devoted themselves to the ivory-tower task of interpreting the classics. The historic centrality of Aristotelian learning in the Western tradition can be explained in part by the strategic role it played in bringing together the argumentative learning of classical Greece and delivering it to this new breed of scholars and scholarship that emerged in the Hellenist period (Africa, 1968).

The triumph of the Confucian paradigm in China was even more directly influenced by political factors. Political intervention was facilitated after the Warring States period by the establishment of the powerful, centralized Ch'in-Han state and the growth of an extensive bureaucratic system, both of which made thought control a real possibility—particularly in the wake of the first Ch'in emperor's infamous "burning of the books and burying of scholars." Yet the facts are somewhat more complicated than the first Ch'in emperor's reputation as a ruthless suppressor of thought might suggest. The sociopolitical writings of the "hundred schools" and the scholarly books found in private collections were all destroyed, but the imperial decree did not include the emperor's own holdings, nor the libraries of scholars in court employ, which, in effect, received court protection and survived into the Han (Kaizuka, 1967).

The Confucius to whom these scholars looked as their progenitor had little idea that he was engaged in what his later followers would call learning (Tsuda, 1948). But the culture that Confucius had extolled came to be regarded as the scholarly ideal, and Confucius was promoted as the ideal scholar (Hiraoka, 1946). In Mencius (early 3rd century B.C.) there are still vestiges of the older wandering scholars, but with Hsun T'zu (later in the same century) the style has become more academic and Confucianism has begun to acquire a canonic form. At this juncture Confucius the historical person gives way in the scholarly mind to Confucius the symbolic conception.

As a proponent of the study of man as distinct from nature, Confucius has often been compared with Socrates. Had the latter's disciples codified his teachings and shaped them into a fixed body of doctrine, it is at least conceivable that they would have come to occupy a dominant position in the Western tradition, giving it an exclusively humanist bent similar to that of Confucianism. In any event, for those who would view science history as a linear unfolding, Socrates's rebuff of the "empty theories" of the Ionian natural philosophers and his efforts to return men's attention to the human world should cast him in a reactionary role (Levin, 1948).

What is the well-lived life? Sensitive people everywhere continue to seek an answer to this basic human question. At the times when it seems most pressing, questions about nature (*physis*)—about the world external to and divorced from man—appear trivial, superficial, and even somewhat childish. Yet this question about the meaning of life was asked long before the advent of science and continues to be raised today. We have little sense that cumulative progress has been made, little confidence that we have any better answers than the sages of old. Thus, whether or not virtue can be taught, as Socrates and his followers believed, our knowledge of it does not exhibit cumulative progress. Questions about the nature of man are never resolved. Indeed, it is only because historical changes create new human ecologies that new questions have been generated at all. In this sense moral knowledge is not science (*scientia*) but wisdom (*sapientia*). Science is knowledge that can be accumulated through normal science. The problems that spring from the moralist's quest, however basic, must be answered by each individual for himself. The traditional concerns of the moralist have never been successfully transformed into normal science, as a quest for *scientia,* and one can think of few things that would lead more quickly to the decline of the questing spirit itself. History affords many cases in which men have tended, in the course of repeatedly asking the same basic questions, to lose their enthusiasm and fall into moralizing and didacticism. This type of sterility can frequently be seen in the Confucian tradition, giving credence to the argument that it was only when basic problems of human nature were abandoned as finally unresolvable that the road to the demonstrative, cumulative progress of modern science was opened.

But if Confucianism contained little that showed promise of development into modern science, who were the Confucianists and what did they do? Although the Confucian classics are attributed to Confucius, he himself probably did not compile them. He was simply the preeminent embodiment of the attitude with which it was felt the classics should be approached. If this attitude can be said to have given rise to a normal-science-like development, it would be the exegetical study of the Four Books and Five Classics. That is, the most important function of the Confucian scholar was to be a compiler of ancient texts. Practically speaking, little more seemed necessary, for when emperor Han Wu Ti (154–87 B.C.) accepted Tung Chung-shu's proposal that "all that is not encompassed by the arts and disciplines of Confucius be prohibited," Confucianism became orthodoxy and the Hundred Schools disappeared from the stage as state-supported actors.

As they emerged from the hands of the Han Confucianists, the Five Classics (the *Book of Odes,* the *Book of Documents,* the *Book of Rites,* the *Book of Changes,* and the *Spring and Autumn Annals*) were all-encompassing works, amorphous in a way that embraced rather than excluded or denounced heterodox doctrines. Commenting on this characteristic, Ssu-ma Ch'ien's father, Ssu-ma T'an, once observed that "the Confucianists were too broad and lacked a vital center."[6] Yet it was precisely this "breadth without edges" that enabled them to survive the Ch'in-Han consolidation and unification of Chinese thought. Doctrines of the Yin-yang theorists and others outside Confucianism proper found their way into the Confucian texts, but orthodox scholarship came to be conducted in an atmosphere that reflected its progenitor's lack of interest in the scientific dimension. The subsequent course of Chinese scholarship had now been set.

With the advent of the Chinese bureaucracy and its examina-

[6] See the "Preface of the Grand Historian" to the *Shih Chi* [*The Historical Records*]. In English see *Records of the Grand Historian of China*, tr. Burton Watson, 2 vols. (1961).

tion system, learning clearly became the province of a gentry class that aspired to office and the vocation of governing. But in seeking to purvey knowledge to the rulers of various principalities, the wandering scholar of the Hundred Schools period equally desired a government post. Thus, the art of governance was the main burden of his teaching. There was little opportunity for him to establish himself as an independent professional, and little likelihood that he could have made a living by teaching the citizenry as men with scholarly interests did in classical Greece. His predicament in this respect can perhaps be attributed to a difference between Oriental hierarchy and the entrepreneurial spirit of the West—a difference that is reflected even in things like astrology. While astrology in China continued to focus on divining the significance of extraordinary celestial phenomena for the Son of Heaven, the Hellenic-Hellenist West witnessed the growth of a horoscopic astrology that was supported by and concerned itself with the fate of private patrons who hired astrologers to read their stars.

Chinese rhetorical learning and political discourse was thus more oriented toward government service than was the case in the West. When men nurtured in this tradition were confronted with the choice of a paradigmatic personage from among the "hundred philosophers," they made their selection with an eye to acquiring a government post and the opportunities for fame and fortune that went with it. Meanwhile, with its emphasis on social order and its close connections at the Han court, Confucian studies received official recognition, and Confucian scholars were made government officials. Henceforth men who sought distinction and prosperity through learning would flock to the Confucian fold. In the early Han period this learning was largely instruction in government for the edification of the Emperor. With the growth of a centralized bureaucratic order, the official character and orientation of Confucian scholarship grew even stronger.

If Confucianism proved particularly attractive to those who chose a single paradigm from among the Hundred Schools, this was not necessarily due to the logical coherence or objective truth of its doctrines. Had the matter been left to verbal disputation, the Legalists might have had the upper hand. In this sense, the triumph of Confucianism in the early Han vividly illustrates the proposition that politics and other extraneous factors are at times deeply involved in the process of paradigm selection.

At first glance the emergence of the Aristotelian paradigm in the West presents a sharply contrasting picture. Aristotle's doctrines were logically persuasive, for in presenting them he took up, criticized, and rejected earlier opinions one by one. Yet, other factors contributed to his eventual triumph. The anti-hierarchical attitude embodied in Democritus' atomic theory could not have endeared him to Hellenist regimes, and his views would have met with even greater resistance from the Islamic society of a later period and the theology of the medieval West. A difference in degree is unmistakable, but it was the same broad-minded tolerance for a range of opinion that is evident in the Confucian case that enabled Aristotelian thought to make its peace with the established order.

A Note on Objectivity

A court of law seeks objective standards by which to measure the justice of human actions, but for scientific communities confronted with decisions about the value of a new piece of research, potential interpretative significance is a prior consideration. To suggest that scientists employ criteria of their own making in these decisions may sound paradoxical, for science has long been identified with the quest for objective truth. Meeting objective standards is, to be sure, an effective means of securing support for one's theories. Yet throughout the history of science such "objective standards" have changed time and again. When they are the standards necessary at any given time to ensure that the results of academic research is smoothly transmitted to the next generation of scholars or made readily available for other, non-academic purposes, those on the frontiers of scholarship tend to march to a different drummer. $1 + 1 = 2$ may be an objectively sanctioned proposition, but this does not make it stimulating to the research appetite. The researcher's interest is much less likely to be aroused by the question of "objectivity" than by a problem of choice—such as what sort of paradigm will better serve to generate intriguing problems and open up new areas and dimensions of research.

The advent of Copernicus' heliocentric hypothesis presented

the scientific community with just such a choice, for the movements of celestial bodies were already quite explicable in terms of Ptolemaic geocentric theory. The Copernican hypothesis did reveal an order and a system that had been hidden in Ptolemy's planetary models, but the advantages were not all on its side. Computer calculations have recently suggested that, on the whole, contemporary Ptolemaic theory actually gave a better account of the heavens as then observed. Ideally, the paradigmatic achievement which leads to a scientific revolution rests on a solid base, but new paradigms are seldom made to order. With occasional exceptions, the new achievement becomes a firmly grounded paradigm only as it is developed and articulated by later generations. The author of the heliocentric hypothesis had no means of physically demonstrating his hypothesis. In fact, Copernicus was not a first-rate observer, and his writings do not offer much substantive support for his views. Thus when Galileo and Kepler became fascinated with and opted for the heliocentric theory, they were not responding to "objective considerations." Their conversion to the Copernican hypothesis stemmed mainly from their discernment of its potential for future research. "Positivistic" astronomers, however, did not accept it. Indeed, for scholars who felt obliged to weigh the matter in terms of objective evidence, then available data would seem to have dictated rejection of the Copernican hypothesis. Discovery of the annual parallax of the fixed stars would have won support for the Copernican view (see Figure 2–3), but the parallax was an extremely small value that

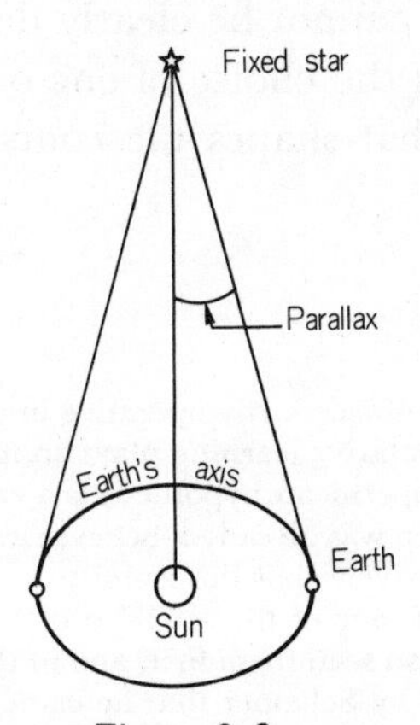

Figure 2–3

could not be detected with the naked eye. Unable to find it, even a first-rate observer like Tycho Brahe did not adopt the Copernican hypothesis. Ninety-five percent of the current books on astronomy opposed his theory, and if the matter had been left to prevailing academic opinion, Copernican theory might well have been buried and forgotten. After Galileo's telescope had revealed the satellites of Jupiter and the waxing and waning of Venus, he did seek to use these discoveries to bolster the heliocentric hypothesis, but this was because these phenomena discredited the Aristotelian cosmos, not because they actually proved the Copernican position.[7]

By the late 17th century, however, scholarly opinion had gravitated toward the Copernican hypothesis, and most scholars had adopted the heliocentric paradigm even though it had yet to be adequately verified. Having taken the Copernican position, they set about to improve the observational precision of the telescope, determined to discover the parallax that would yield verification. This discovery was finally made in the 19th century, but in the interim, the search for the parallax was the greatest challenge to the early modern astronomer. It was the choice of the Copernican paradigm, and not conclusive proof of it, that sustained the research of generations of astronomers.

The realm of technology provides even more vivid examples of this kind of choice and its consequences. Whether the electric engine is preferable to the internal combustion engine, whether a rotary engine will prove more desirable than a reciprocal one—these are things that cannot be clearly determined until each has been developed. It is the choice of one over the other at the predevelopment stage that shapes the course of subsequent technological history.

[7] Rational criteria are not necessarily operative in paradigm selection. The willfulness so common in argumentative learning plays an important role in these decisions. Galileo's support of the Copernican hypothesis is a case in point. Although the standard view holds that Galileo was already a believer in Copernicus at an early age, W. Hartner has argued persuasively that this is not true. Hartner claims that it was only after Galileo's discovery of four of the satellites of Jupiter led to a dispute with the Jesuit Scheiner over who had seen them first, and in the course of opposing the system of Tycho Brahe supported by Scheiner that he came to side clearly with the Copernican view.

Codification

Prior to the articulation of paradigms, individual teachers are identified with the learning they profess. The general attitudes and ways of thinking of Socrates and Confucius were manifest not in systematic treatises and discourses but in the dialogues and debates that were part and parcel of rhetorical learning. Until they became identifiable as distinct scholarly traditions, and until they came to be accepted by a given group of scholars as particular modes of learning, their teachings were without a name. What could one call them—Socratology and Confucianity? The immediate disciples of these two men received and interpreted their master's doctrines orally and responded to particular problems and situations by asking what he would have thought about them. For them, the master himself was the paradigm: personality and doctrine had not as yet been differentiated.

At this stage, the naked voice is the most important means of communication, recourse to the written word being taken when no other alternative is available. Yet by the time one comes to the next generation, to the disciples who have had no direct contact with the Master, authority begins to be depersonalized. The learning of Aristotle is divorced from his person and becomes Aristotelian learning, the proper name of a specific body of knowledge. And in the course of its refinement as a scholarly paradigm, this body of knowledge assumes its canonical, written form. Likewise, the *Analects* of Confucius is a compilation by disciples.

In the West, this codification of learning reached its peak in Hellenist Greece during the 3rd–2nd centuries B.C., a period epitomized by the work of Euclid and Hipparchos and sometimes referred to as "the age of textbooks." It was in this period that the several branches of ancient science received systematic expression. The Alexandrian museum, where much of this work was carried out, is said to have housed 500,000 volumes at this time—a significant collection even though two zeros must be dropped from this figure when comparing these papyrus manuscripts with the printed books in modern libraries.

The emergent doxographical tradition is another feature of such times. Carrying on the work of Theophrastus, who had himself fully assimilated and systematized the learning of Aristotle, this school focused its efforts on bringing together, classifying, and annotating fragments of classical texts in a manner similar to the exegetical scholarship of Han and T'ang China. After Andronicus Rhodius (fl. 70 B.C.), the school concentrated exclusively on annotating Aristotle, and before the end of the Hellenist period, classic texts had come into being in many fields, from Euclid's *Elements of Geometry* to Ptolemy's *Almagest*.

In China, the systematization of learning that had begun in the Warring States period was initially completed in the Han, forming the prototype of subsequent Chinese scholarship. Groups of Confucianists had earlier coalesced into an identifiable school, and as they found themselves pitted against other schools, the *Book of Documents* (*Shu ching*) and the *Book of Odes* (*Shih ching*) had been brought together as a kind of teaching canon. But it was in the Han period that Confucianism became orthodox doctrine and the canonical position of the five (or six) Confucian classics became firmly established. Classics in the fields of astronomy (the *T'ien-kuan shu* chapter of the *Shih chi*), mathematics (the *Chiu-chang suan-shu*), medicine (the *Huang-ti nei-ching* and the *Shang-han-lun*) and pharmacology (the *Pen-ts'ao kang-mu*) were also completed at this time; and with the appearance of Ssu-ma Ch'ien's *Shih chi*, China's official dynastic histories acquired their paradigmatic form. The Han library classification system was a by-product of this codification process. In this system, the Six Arts of Confucius occupied a central position as classics, while the teachings of the Hundred Philosophers were all lumped together under the rubric "other schools." It is these classic texts of the Hellenistic and Han eras that I had most immediately in mind when I used the word "paradigm" to refer to "canonical codifications of classical texts which set scholarly style, legitimate the specialized, professional activities of intellectual groups, and lay out a course of subsequent development for what has come to be normatively defined as scholarly activity."

Once learning achieved a paradigmatic form, once the founder's teachings had been codified by an advocate group, scholarly practice underwent a radical transformation. Mathematics became Euclidean in the West, while it followed the style of the *Chiu-chang suan-shu* in China. No other ways of doing mathematics were now considered legitimate. In practical terms this meant that the

Western follower of Euclid was henceforth to begin with axioms and solve his problems with a ruler and a compass, while one who did his math in the Chinese tradition need no longer worry too much about proofs, it being considered sufficient simply to show the problem and the answer. In astronomy, geometric methods became standard in the West, while algebra was favored in the East. In practice the Chinese also used geometric figures in solving astronomical problems, but since such figures did not appear in the mathematical canons, they were not employed in orthodox calendrical writings.

Decisive differences of focus, scope, and purpose also flowed from the codification of ancient learning. In the West, the resolution of a problem in planetary motion ultimately involved a discussion of the nature of the universe, but this problem was only of tangential concern in the East where the main goal of astronomy was the prediction of solar and lunar eclipses. So deeply ingrained was the Chinese orientation that one T'ang astronomer flatly asserted: "Disputing about the nature of the cosmos is not our business." Their concern, and the chief task of generation after generation of Chinese astronomers who patterned their work on the *T'ien-kuan shu (Shih chi)* and the *T'ien-wen chih* (*Han shu*) was compiling records of extraordinary heavenly occurrences—an activity that was unimportant in the West.

The story is told of a simpleton who, upon hearing that he was in the presence of an economist, asked him the price of cognac. The economist, it turned out, did not know—not because he preferred beer, but because professional economists have considered such problems peripheral to their science ever since economics became an academic discipline. Since the standardization of scholarly practice, a scholar can be regarded as an astronomer without knowing the names of the stars, or as a botanist without familiarity with the appellations of trees and plants. Paradigm codification produces this kind of academic style that permits the scholar to focus his attention on certain clearly delineated concerns and forget about "extraneous" matters. Methodology as well as subject matter becomes restricted, and some phenomena, now considered "unacademic," disappear completely from the scholar's purview.

A Note on the Meaning of the Classics

The ancient practitioners of rhetorical learning had been chiefly concerned with their own thought and its oral expression. Generally speaking, they had not considered written materials particularly important. With the codification of ancient learning, scholarship came to mean the study of books. Even so, the classical texts did not come into being overnight. In their mature form both the Aristotelian corpus and the Confucian classics were the product of men who added their interpretations to the utterances of their mentors. Classics in the fields of medicine and materia medica regularly incorporated large amounts of new data within the old paradigmatic categories. How is this phenomenon to be explained? Why did ancient scholarship find in the annotation and exposition of the classics its peculiar pattern of growth and development?

Science is *avant-garde* thought. Forever destroying fixed ideas and settled opinions, it marches incessantly onward, acknowledging no authority save that of nature. So goes the heroic view. But science does not actually advance in this way. In fact, no coherent results are likely to emerge from the endless round of construction and destruction implicit in this view. Put more positively, science seeks and constructs authoritative paradigms. If ancient scholars found these paradigms in classical texts it was because there were no alternatives. Prior to the advent of printing, books were scarce, and anyone wishing to study ordinarily had to begin by copying (or, in China, by memorizing) the works of predecessors. Much of the young scholar's energy was thus inevitably spent in transcribing, comparing, and emending texts. Moreover, when the manuscript is the sole medium of transmission, the distinction between author and reader is easily lost. Reading and copying occur together; and when the reader scribbles his own opinions in the margins or skips over portions of the text that are of no interest to him, he has in effect become an annotator and interpreter, the "editor" of a manuscript that is a commentary in its own right (Rosenthal, 1947).

Some works attributed to Aristotle and Confucius could only have been intentionally fabricated by later generations. A deliberate attempt to disguise one's scholarship seems strange in today's world where nothing counts so much as tangible achievements and the scholar is always anxious to get his own name in print. But when books were transcribed by hand, pretending to the name of an authoritative figure was the normal means of linking one's work to a tradition. It did not carry with it the stigma attached to forgery today.

The practice was part of the larger pattern of manuscript-centered, classics-oriented scholarship, in which the activities of individuals had to be assimilated to an authoritative tradition. Footnoting, in which a writer cites the authors and works used, makes sense in a world of printed books: a reference to such-and-such a page of so-and-so's work is immediately understood, and the book itself is often readily available. In an age of manuscripts, however, citation was hampered by the lack of common pagination. Moreover, even if scholars had quoted other annotators and lesser emenders of the classical texts, manuscripts would have been so difficult to obtain that such references would have been largely useless. And yet, scholarly discourse and discussion cannot proceed without some kind of communal property. In attributing their work to the founder of the tradition, the heirs of men like Aristotle and Confucius found a convenient means of meeting this need. At the same time this procedure served to enhance the authority of the founder's achievement and hence of those who had taken it as their paradigm. In this context the question of authorship was not a significant issue. The name of the progenitor and the texts associated with him were used first of all as an invisible yet commonly understood badge of identity by which scholars knew themselves and were known to others.

But "accumulation" and "progress" are evident even in this pattern of scholarship. Adding a brick of newly acquired knowledge to the previously constructed paradigmatic edifice within and from which one works is, at the very least, more economical than endless hours spent in controversy and disputation.

In modern experimental science, radically new observations and experimental devices occupy a status fully commensurate with paradigms set down in books, and the classics have yielded place to widely disseminated printed textbooks, journal articles, and handbooks. These modern-day "classics" are designed to bring successors abreast of current standards in normal science in the shortest possible time. But as a vehicle for the efficient acquisition of existing knowledge, the textbook has little room for accounts of the painstaking labor that scientific research requires, and of the circuitous and often tortuous paths that it takes. However functional, however clear and efficient for teaching the student how to set up problems and work out solutions, the wholly predictable style of contemporary scientific treatises, textbooks, and handbooks does little to provoke creative thought. (For a discussion of the development of modern academic journals, see Chapter 4.) But even in classical times, the astronomical charts of ancient Babylon and Islam, and the Chinese astronomical treatises were little more than a listing of dry-as-dust constants and results. Much more schematic than today's scientific treatises, they manifest no skepticism at all about the principles and rules underlying their construction.

Nevertheless, the value of the classics is being reasserted today because learning has proceeded on its present course for so long and has come to be transmitted in a manner so routine, general, and bookish that we tend to lose sight of its foundations. In many fields, classic texts offer an opportunity to become reacquainted with these foundations, and permit exploration of the variety of ideas and modes of expression that prevailed prior to the development of a paradigm into a normal science tradition. What has become paradigmatic for normal science is only a selected portion of the original achievement; as the history of Neo-Confucianism (in both its Chu Hsi and Wang Yang-ming forms) suggests, emphasis on aspects of the paradigm that were neglected in the original selection can sometimes lead to the development of a distinctly different mode of learning. In returning to the classic text, then, the reader can still come in contact with the mind of the founder, with an individuality that was lost in the attempt to make of his achievement a complete and universal paradigm. This encounter may prove to be a valuable stimulus to thought. At a time of major disciplinary reorientation, the classics sometimes have meaning even in fields that have gone very far in the direction of normal science. Thus, though in one sense a classic can be said to die once it has given birth to a normal science tradition, it continues to have life as long as scholars find themselves stimulated by it.

The Acquisition of an Institutional Base

In putting together a new religious sect four things are necessary: 1) a founder; 2) believers; 3) doctrine and scriptures; 4) a believers' organization and physical facilities of some kind. To liken groups of contemporary scientific specialists to religious associations is, I suppose, quite inappropriate, but scholarly groups in the ancient world—from the Pythagoreans in Greece to the Confucianists and Mohists in China—often had a distinctly religious flavor. The Hippocratic school, the Platonist Academics, and the Peripatetic followers of Aristotle also bore a strong resemblance to religious collectives both in their deportment and in their development as scholarly groups. The movement toward the codification of learning can be seen, at least in part, as an expression of the desire of these and other advocate groups for a "rule" or "order."

When advocate groups cease to be merely groups of believers and become professionals, then the social history of a paradigm has entered a new phase—institutionalization. A professional group must have some sort of institutional base, and, conversely, it becomes viable as a professional group once it has acquired that base. In order to win a place for themselves in the larger society, advocate groups commonly seek out the wielders of power and influence, and proclaim the originality and virtues of their tradition. But in late-developing countries which simply import learning from abroad, the development of institutional structures occurs first and professional groups are artificially nurtured within the framework of these structures (see Chapter 6).

For the scholar, the most important consequence of professionalization is that he can now make a living by his scholarship. As a result, he is likely tempted to see the institution that sustains him more as a source of livelihood than as an instrumentality for education and research. Three common types of professional groups can be distinguished: 1) the independents; 2) employees in the bureaucratic structures of government; and 3) employees in educational institutions. The latter group is of primary interest, for it is finally through installation in institutions of higher education that an academic tradition rooted in a particular paradigm secures its perpetuation. (Chapter 5 will deal with this question in more detail.)

Again, the history of Marxism is a classic example of this process. First there are the paradigmatic figures Marx and Engels. Next come advocate groups who publish editions of their complete works, codifying and interpreting their achievements. Finally a chair in Marxist economics is introduced into the educational system. With this acquisition of an institutional mechanism for the production of heirs, for the training of those who will continue to do scholarship in the Marxian mode, the process by which a revolutionary paradigm becomes conventional is complete.

The process by which scholarly groups find a niche in a social institution stands in need of more detailed research, but a few general observations can be made. Inasmuch as academic paradigms often evolve outside the confines of established institutions, the attempt of advocate groups to secure a permanent place for themselves within an established institution almost inevitably gives rise to tension between the group and incumbent defenders of the status quo. Not infrequently, the aspiring tradition is absorbed by the system, subordinated to the structures of power, and intellectually domesticated, losing its creativity in the process. Even in fields and modes of learning with little apparent ideological significance, the efforts of scholarly adherents to come to terms with an establishment sponsor have sometimes radically altered the course of their subsequent development and cast a long shadow over the very meaning of scholarship and scholarly activity itself. A glance at the type of learning that took root in the medieval universities of the West and that which grew up in conjunction with the Chinese examination system should make this quite clear (see Chapter 3).

The Contours of Greek Scholarship

Originating with a single teacher whose doctrines were carried on by successive generations of "masters," the Greek schools—and particularly the philosophic bands—were strongly fraternal in character. In the Hellenist period Aristotelian dominance was still a thing of the future, only one of four main schools (Platonists, Aristotelians, Stoics, and Cynics) that sustained their traditions unbroken until the end of the ancient world. On the periphery, of course, there were other schools and patterns of scholarly life. Some itinerant Cynics and Stoics continued to expound their philosophical principles on street corners; by the 3rd and 4th centuries independent schools had emerged in several regions. In keeping with the Greek tradition, these schools regularly engaged in heated controversies with one another, but within the religio-fraternal framework of each group the work of codification proceeded steadily. Courses of study were devised for the training of successors, and scholarship began to make itself at home in an institutional setting.

The center of this activity, as of Hellenist science in general, was the Museum at Alexandria. Universally regarded as the first research facility of its kind, this government institute attracted poets, writers, critics, geometers, doctors, astronomers, historians, and other scholars, who lived together in separate quarters near the palace and received official stipends. Said to have numbered as many as a hundred during the reign of Ptolemy II (285–246 B.C.), this community of scholars engaged in wide-ranging debates and discussions that transcended sectarian lines. The institutionalized research setting provided by the Museum proved particularly conducive to progress in philological studies.

To a degree that might have surprised their classical counterparts, the Greek scholars of the Hellenist period practiced the arts of caution and subservience. Public support of political discourse had disappeared, Democrats and Cynics were not welcome, and those who showed signs of disloyalty to the regime were subject to purge. Yet, when Greek scholarship retreated from the political arena, it found expression in a flowering of science. The age of Ptolemaic despotism was also the age of Euclid and the astronomers Aristarchus and Eratosthenes. Precision improved most markedly in observational astronomy. Government-aided research in military technology bore noted results in the work of Archimedes and Heron, and experimental methods foreshadowed the 17th century's mechanistic approach to nature (Africa, 1968: 60).

Important changes also took place in the crucial areas of education and curriculum development. As the map of the Greek world expanded, the education of the classical period, centered in private academies, broadened. Increased numbers received schooling, and education began to take on a rather public character, even when conducted under private auspices. Textbooks and lectures replaced face-to-face dialogue, and education came to be more and more formally organized. Residential, middle-level schools (*ephebeia*) were built throughout the Mediterranean world for the children of the upper classes.

The seven liberal arts (the trivium: grammar, rhetoric, and logic; and the quadrivium: arithmetic, geometry, astronomy, and harmonics) that were to form the backbone of Western education were definitively formulated in the middle of the first century before Christ. Dialectics, philosophy, and medicine, on the other hand, became specialized subjects to be pursued at a higher level. (The dialectic was converted into jurisprudence at the hands of the Romans who established higher schools of law, while philosophy was annexed to Christianity and transformed into theology. It was in this form that law and philosophy would eventually appear, together with medicine, as the three professional faculties of the medieval university.)

Hellenistic culture has been described as a "lecturers' culture," its typical form being the formal public speech (Marrou, 1956). Although lectures seem to have been given earlier at Aristotle's Lyceum, it was in the Hellenist period that the lecture form was perfected as a method of instruction. When teaching in schools, even astronomers and physicians were expected to be good lecturers who could expound on the classics with some eloquence. As a result, oratorical skills and the trivium and dialectic in which they were learned enjoyed such prestige that these offspring of the classical argumentative tradition were considered basic for everyone. Medicine seems to have been carried on from the beginning in relatively specialized, guild-like groups, and dialectical skills that might be used to contest the clinical practices of other groups became an essential part of the accomplished physician's baggage. Galen, that sage among physicians, had Sophistic training.

Plato's mathematical quadrivium, the other classical component of the Western educational tradition, did not fare so well during this period. In the *Republic*, Plato had proposed that before men enter into philosophical studies at age 30 they should prepare themselves with a decade of training in the several branches of mathematics. Ultimately Nicomachus' *Introduction to Arithmetic* (ca. 1st century B.C.), Euclid's *Elements of Geometry*, the work of Geminus (*Introduction to Phenomena*) and the popular poem of Aratus (*Phaenomena*) in astronomy, and Aristoxenus's *Elements of Rhythm* served as textbooks for these studies.

But Plato's admonition was not always strictly observed. The Epicureans and Skeptics had little use for mathematics. As mathematical subjects disappeared from the curriculum teachers became so scarce that even Platonists could only encourage students to study it on their own. Wherever mathematics was preserved, it had little or no relation to the Neo-Platonist philosophy of the late Hellenist period. Students began their studies with the Ionian natural philosophers; much like students today, however, they did not read the original works but relied on secondary, copied textbooks.

Meanwhile, the achievements of such figures as Plato, Aristotle, Epicurus, and Zeno appeared in canonical versions, and philosophy itself seemed in danger of degeneration into exegesis. Plato's educational program was eventually overwhelmed by a classics-oriented emphasis on the trivium that drove the mathematical sciences from middle- to higher-level education and transformed them into the work of specialists. By the late Hellenist period, government had become directly involved in the licensing of teachers and in the organization of education, and the creative spirit withered away.

The Institutionalization of Learning in China

A curriculum consisting of the Six Arts (Ritual, Music, Archery, Horsemanship, Letters, and Arithmetic Computations) is mentioned in Chou regulations, but only during the two Han eras did a clearly defined system of schooling emerge. The crucial event took place under the Han Emperor, Wu Ti, who established formal lectureships in Confucian studies and provided government stipends for scholarly experts on the Five Confucian Classics. The education over which these erudites presided was specifically aimed at the training and development of administrative officials. Teachers themselves were members of the bureaucracy, and regularly moved back and forth between the educational and administrative departments of government.

Study of the Confucian classics stood at the center of what was in effect a new curriculum. Preservation of the ceremonial traditions associated with the study of music and ritual, archery and horsemanship now became an administrative function, while the arts of calligraphy and computation became adjuncts. The scholarship of mathematicians, astronomers, physicians, and other technical specialists was distinctly subordinate.

That the Confucian classics outlived the other achievements of the Warring States period cannot be adequately explained by the free choice of any one generation. It was almost wholly the result of an official policy which accepted Tung Chung-shu's recommendation that the canon of official learning be restricted. This policy made the ideal prince-subject relationship a key instrument of political control. Before a pluralistic intellectual tradition that had found expression in the doctrines of a hundred schools was able to consolidate itself as a coherent paradigm, government intervention set learning on a more restricted course and then cast it in that most durable of all political molds—the Chinese bureaucratic order.

If the development of Chinese learning seems in retrospect to have had a tragic dimension, the beginnings of that tragedy are to be found here in Han times. For the "Confucian" canon that was brought together within this matrix included much more than the rites and music of the court Confucianists. The study of Yin-yang, natural disasters, and even the Taoist thought of Lao-tzu and Chuang-tzu were included in the Confucian medicine chest that now held a variety of intellectual perspectives and socio-political visions for which the Six Classics merely provided a general foundation. In view of the cut-and-dried tone that prevailed in this mixed bag of doctrines and documents, one wonders whether the classics were ever of much use. Perhaps they were useful precisely because they contained very little that was "alive" or relevant to the practical realities with which officials had to deal. Their use was in quelling the disputes of the hundred schools and bringing a kind of intellectual uniformity to the realm.

In the Sui (581–618), these Confucian classics were explicitly designated as the basis of a revived civil service examination system; and in the T'ang, the examination and educational systems were linked in a more systematic fashion. T'ang statutes prescribed training in the Confucian classics (the *Analects*, etc.) and the histories (the *Shih chi*, *Han shu*, etc.) both in schools and for appointment to office. Classic texts were also designated in such specialized studies as law (the penal, *lü*, and the administrative, *ling*, codes), arithmetic (the *Chiu-chang suan-shu*), and astronomy (the *Chou-pi suan-ching*). Designation of the classics (in their orthodox interpretations) as a kind of official textbook to be studied in preparation for the civil service examinations rounded out the system of official education. Buddhist texts, it may be observed, were never used as textbooks in schools. If scholars did occasionally lecture on them, it was simply to criticize and reject them.

Bureaucrats are ever mindful of precedent, and scholarship based in a bureaucratic order tends to judge a new mode of thought or scholarly practice less for its future potential than for its antecedents. The older the precedent, the better, for age itself is regarded as evidence that what now appears "new" has stood the test of time and survived the trials of history. In China this bureaucratic respect for the classics seeped even into fields like astronomy as men unreservedly put their faith in dubious records of the pre-Ch'in period and strove to revise theory to accommodate old data.

Common Notions of Learning in East and West

The scholarly disciplines that were henceforth to be regarded as learning had already settled into their respective discernible patterns by the Hellenist/Han period. In the West, philosophy, medicine, the dialectic, and the mathematical sciences (including astronomy and statics) were considered standard academic pursuits, while inquiries into chemical properties were excluded from the ranks of legitimate learning. In China, study of the Confucian canon occupied the highest position, while astrology and calendar-making played a supporting role. Medicine ranked far down the list, and the status of mathematics was even lower. Technology and applied science did not find favor in either world. Nothing resembling these activities can be found in the Platonic curriculum, nor were they recognized as higher learning in China.

Underlying these two configurations, more basic differences can be detected. In Western tradition, the influence of rhetoric and logic was pervasive, whereas Chinese scholarship was dominated by documentary studies that sought to observe and preserve the traditional conventions of the histories and the classics. The Chinese tradition's emphasis on the development of a class of scholar-officials who could protect and defend orthodox doctrine in a cautious, cool-headed, and pragmatic manner also sets it apart from the Hellenistic type of training that produced Sophist rhetoricians with an argument for every occasion. But even more striking is the contrast between the esteem with which the Platonic mathematical tradition was regarded in the West and the Chinese preference for treating mathematics as a minor technique to be practiced by lower-level bureaucrats. Insofar as the emergence of modern science presupposed the revival of Platonism, it is possible to argue that this divergence in attitudes was, in the long run, a crucial one (see Chapter 4). Yet, as has often been observed by those who prefer to trace the origins of the Scientific Revolution to the Aristotelian-Scholastic tradition, the Platonic revival was a Renaissance phenomenon. From their perspective, the 13th-century revival of learning appears more important, the existence of a strong logical and rhetorical tradition a more fundamental watershed.

We shall return to this question later, but the major outlines of our approach can be stated here. Every society may be said to have its own view of learning—first of all in widely held notions of what scholarship is and what it means to pursue it. In China, mention of *hsueh-wen* immediately called to mind what one did to prepare for examinations, whereas scholarly activity in the West tended to be identified with persuasive discourse and discussion. Although these notions are often unarticulated, exploration of a given culture during a given period typically uncovers something quite specific—a particular mode of learning that serves as a kind of "model discipline." The "model discipline" is in its time the queen of the sciences, envied and imitated on all sides by those who would share its prestige.

We must pause to ask, however, by what virtue particular modes of learning come to occupy this status. One important point is that they are autonomous activities with their own goals, disciplines not subjected to or dependent upon other disciplines or non-academic values. Historically speaking, however, this condition seems to have been met in two different ways: 1) by modes of learning whose methodology has been widely used as a paradigm by scholars in other fields, and 2) by disciplines that enjoy a distinguished reputation and high social standing. Euclid's theorems, mechanics in early modern science, and physics in a later period may be considered methodologies of the first kind. Examples of the second can be found in the ancient and medieval astronomy of the West and the classical study of China.

Application of this notion of a "model discipline" to the two major academic configurations outlined above allows us to see them not simply as static hierarchies but as dynamic patterns that operated across rather large expanses of time to shape the growth and development of scholarship. Moreover, in viewing the rhetoric and logic of the West and the Confucian studies of China as "models" rather than as merely primary disciplines, a contrast in the general character of the two configurations emerges. In the West, the model disciplines exercised extensive methodological influence, whereas the scholarly status of Confucian studies in China was simply a function of their social prestige. Confucian studies did develop an exegetical method of sorts, and Confucianists themselves were preoccupied with their own scholarship—looking upon mathematics, astronomy, medicine, and the other "instrumental" studies as petty arts; but practitioners of these arts were seldom directly influenced by Confucian methodology, even though they continued to feel that their knowledge and skills were somehow inferior to the noble arts of Confucian learning. In the broadest terms, it is at this point that the decisive difference between the academic traditions of East and West is to be found.

An Orderly Universe? Analysis vs. Classification

Despite the fact that the ancient practitioners of rhetorical learning often appeared to be preoccupied with their combative fortunes, and even though the types of questions with which they dealt are seldom if ever resolvable, the format of their encounter called for a determination of winners and losers that both presupposed and engendered a quest for the one "correct" conclusion. Documentary learning, on the other hand, considered its task complete when everything at hand had been recorded. If the rhetorical learning of the West can be said to have yielded a single, unitary paradigm, this paradigm was to serve as the absolute truth and standard of judgment in all matters. Though it was the Aristotelian paradigm that would eventually be adopted, it was done in a way that embraced the spirit of Plato, a spirit that not only believed in the existence of eternal, unchanging forms behind all things but also held that the world had an underlying regularity of structure.

The notion that the universe can be intelligibly grasped in terms of a single paradigm is, however, simply an article of faith. Nothing assures us *a priori* that the world, or nature, operates in accordance with mathematical laws. Had Europeans been willing to entertain the possibility that heavenly bodies have minds of their own and might therefore occasionally feel like deviating from their usual orbs, they might have been content to let the matter rest. Instead, it was insisted that these heavenly bodies must always follow rigid laws—even that their movements must be expressible in precise mathematical formulas. If one thinks about it, this is a strange conviction. However, from the establishment of the Western paradigm in the Hellenist period to the days of modern science, most Western scholars continued to maintain this unwavering faith in the nomological character of the universe.

The Western observer who looked at the planets from within this tradition and saw that their movements were at variance with the laws of celestial motion would make his observations again, revise laws, or endeavor in some other way to bring the laws and his observations into harmony. In this scholarly act, or series of acts, he was insisting that natural phenomena be crammed into a single nomological box. But how did this Westerner react when confronted with something that simply would not fit into the box? How did he respond, for instance, to the sighting of a celestial nova? An event incompatible with the legalistic Aristotelian conviction that the heavens are eternal and unchanging could not be handled within the framework of his intellectual faith. Hence he did not see it. The appearance of a comet drew a similar response, for comets were likewise not considered part of the eternal, unchanging heavens. Could a passing phenomenon in the earth's atmosphere that disappears like a cloud be worth recording? Certain of his answer, the Westerner would not see this phenomenon. This preconceived notion that the heavens were eternal and unchanging directed his attention elsewhere from the very outset.

How did Chinese scholars behave under such circumstances? In China, too, there were modes of learning that concerned themselves with regularities. There was, for instance, the science of calendar-making, whose task it was to observe the movements of the moon and refine the laws of lunar motion. Here an initial effort was made to accommodate observations in the *nomos* box. Yet, when the movements seemed to depart too radically from a normal course, men said that "the moon had lost its motion" or that "the heavens were out of order" and disposed of the problem by placing it in a box marked "extraordinary phenomena." In the *Shih chi,* a separate section was provided for each type of data, and subsequent dynastic histories preserved this tradition. There was no absolutist faith that all the motions of nature could be grasped in terms of mathematical theory. Heavenly bodies move as they will. When human beings have done everything in their power to comprehend them, should they be held accountable for irregularities? If the movements of heaven did not tally with the laws, that was heaven's prerogative.

"Unprejudiced" by a belief that nature was a nomocracy, the Chinese noted the greenness of willows, the crimsonness of flowers—whatever happened to capture their attention—and recorded and classified the data. Because there were, as we have seen, two compartments (calendar-making and astrology) to accommodate celestial phenomena, because the extraordinary was recognized along with the ordinary, records abounded with sightings of celestial novas, comets, atmospheric phenomena like haloes—even some things that we find hard to envisage today such as references to the fixed planets appearing on the face of the moon (coming between the earth and the moon).

Laws of nature are premised on the possibility of repeated verification. Concatenations of historical events do not afford the witness the same opportunity for controlled reinspection, so they resist subjection to laws. Yet from the standpoint of documentary learning it is precisely the peculiar, non-recurrent character of phenomena that makes them worth recording. Interestingly enough, however, modern science has had occasion to be grateful for records of this sort. After World War II, ancient Chinese and Japanese records enabled radio astronomers to confirm that certain spots in the universe found to be emitting powerful radio waves were remnants of novas that had appeared long ago. The historical Chinese distrust of an excessive faith in lawfulness also proved durable. Though a horoscope astrology based on the belief that the fortunes of men (like the movements of the planets) follow a lawful course has been popular in the West since Hellenist times, it was not to prosper in China.

For the Chinese, to engage in scholarship meant to record and classify. Whatever the phenomenon, it was duly noted and put in one of the several compartments set up for classification purposes. Once this had been done, however, the scholar's job was finished. Above and beyond these tasks, he did not plague himself with weighty questions about the lawfulness of what had been observed. Little was challenged, confuted, rejected, or debated; virtually all natural-history-like data were accepted, preserved, and allowed to rest in peace. Generation after generation, large amounts of these data were amassed and arranged in the ample "boxes" provided in the gazetteer sections of the official histories.

At the beginning of this chapter, I described how the appearance of a critical number of anomalous phenomena can provoke a normative crisis within a given scholarly community, a crisis eventually followed by the construction of a new paradigm that ushers in a scientific revolution. The classificatory style of scholarship as it was practiced in China, however, does not give rise to this kind of predicament. In China, the lawful and the anomalous, the regular and the irregular, the ordinary and the extraordinary were all recorded and compartments readied to receive them so that each might find its proper place. If a phenomenon appeared that proved hard to file in one of the existing compartments, a new box could always be made for it. Hence the lawful and the anomalous lived side by side in peaceful coexistence, the system being constructed so as to forestall the emergence of a crisis. In fact, though we have spoken of the anomalous, nothing really can be called abnormal in a context in which it is assumed from the outset that everything should have a legitimate place. Disturbances and convulsions of nature there were, but for the Chinese scholar they were not mysterious or miraculous violations of a law-abiding order. His tradition had equipped him to see them rather as administrative problems. Whatever else it might be, this practice of classifying everything and putting it in a "proper" place is a bureaucratic approach to scholarship.

In the West, where the principle of a single, nomological box was dominant, anomalies were initially ignored. When they became too numerous, however, a crisis arose and a new and more spacious box was constructed to accommodate them. The scientific revolution effected in the shift from Newtonian dynamics to Einstein's theory of relativity is a good example of this process (see Figure 2–4). Propelled by the propensities of an argumentative and logical style, the quest for basic principles found expression in a reductionist impulse and a concern with axiomatic truths that hammered out universal laws and forged binding chains of cause and effect.

In the documentary Chinese tradition, however, classifying and cataloging were paramount. Little interest was shown in the problem of cause and effect so that even when mutual relationships were recognized, no attempt was made to discover a single chain of causes and effects that might link phenomena together consequentially. Attempts to explain everything in terms of nascent principles of natural philosophy—calendrical numbers, the hexagrams of the *I Ching*, Yin-yang, and the Five Elements—had begun to emerge in the controversies of the Warring States period, but in a tradition increasingly dominated by a documentative orientation, their function was attentuated. In pathology and physiology, the Yin-yang and Five Elements theories did serve as the underlying principle of natural philosophy, but in mathematics, calendar-making, and the other quantitative sciences, they became remote abstractions used chiefly to adorn prefaces. At best they provided a rationale for two- and five-fold classificatory schemes.

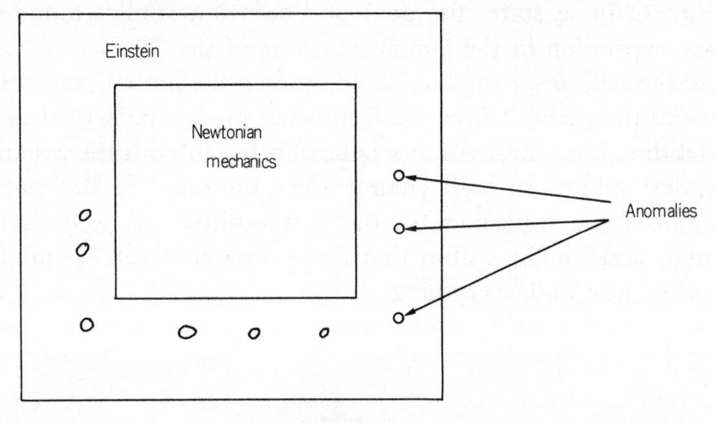

Figure 2–4

The persistent reluctance of the Western academic tradition to acknowledge and accommodate the anomalous might well be seen as its most distinctive feature. A transcendent God ruled over all, and just as men were obliged to obey His will, so nature was constrained to obey absolute laws. The atomists have provided us with perhaps the most radical expression of this attitude. In the atomistic view of nature, the will of man, together with human desires, aesthetic sensibilities, and all other inconstant, unreliable, and hence trivial aspects of life, is devalued, and all sensible phenomena are reduced to matter and motion. Since the atomist would show matter itself to be composed of atoms and void, he is thus able to explain all things in terms of atoms and their motion. No room is left for the intrusion of wayward anomalies.

This working hypothesis (or faith) that all things are obedient to laws stimulated the growth of a natural philosophy, but its influence also extended to reflection on human affairs. Thinkers were persuaded that invariable regularities could be found even in the apparently unpredictable world of politics, and a political science was created to discover them. But should one really expect to find immutable laws behind phenomena as uncertain and impermanent as the political? One is sometimes inclined to feel that in politics irregularity is every bit as real as it seems, but this is not the kind of thought that gives rise to the formulation of a law-oriented science.

For the Chinese, political phenomena were considered worthy of record regardless of whether or not they obeyed rules. But while this led to classification, it did not yield analysis. In the mature Chinese state, the goal of Confucian studies found its classic expression in the famous teaching of the *Ta hsüeh:* "Cultivate the self, order the household, govern the nation, and bring peace to the realm." Like the Confucian mode that served as its model discipline, East Asian scholarship began with the assumption that mutability and change were the ways of the world, recognized the legitimacy of the extraordinary as well as the normal, and sought within that framework to create a suitable place for any and everything.

Chapter 3

Paper, Printing, and the Maturing Traditions

So long as they are nothing more than small movements sustained by a handful of intellectuals, scholarly traditions may be able to retain a reasonable degree of independence. But when a tradition finds institutional expression and becomes part of the social order, extracultural factors in the social substructure exercise a commanding influence. This chapter will focus on a period of several centuries during which learning in its major manifestations acquired settled institutional forms, and during which the differences between the academic traditions of East and West took root within the structures of society at large. Specifically it will examine the Chinese scholarship which established itself within the bureaucratic examination system between the Sui-T'ang and Sung dynasties, and Western learning as found in the Islamic *madrassa* and the European universities of the Middle Ages.

In considering the institutional profile of the two traditions, I shall also be looking at the material media by and through which they were conveyed and communicated. By "material media" I mean, first, the tools and implements of writing and, second, books. No one is likely to contest the assertion that paper and books play a primary role in the transmission of learning. But learning was also transmitted in periods before the advent of printing and in places where paper was lacking. How strongly did the presence or absence of these material artifacts influence the character of scholarship itself?

Scholarly Media and the Chinese Bureaucratic Institution

Bureaucratic government has been called government by documents. Certainly centralized control over a vast imperial state such as China would have been extremely difficult had the authorities been unable to rely on cheap and readily available paper for the promulgation and transmission of official legal codes and directives. In this sense paper played the same kind of role in Imperial China that the telephone and telegraph were to play in the modern state.

The invention of paper is usually attributed to Ts'ai Lun, a Chinese of the later Han, who is thought to have accomplished the feat in about 105 A.D.; some recent sclolarship places the date even earlier than the 2nd century (Pan Chih-sing, 1979). To anyone familiar with traditional forms of Chinese government, it seems a very Chinese invention.[1] Of course the Chinese bureaucratic order antedated paper, and the invention did not, in fact, immediately come into widespread use. Excavations of wooden tablets at Tun-huang and the Imperial Palace in the ancient Japanese capital of Heijō suggest that paper may have remained a fairly precious item for several more centuries. In the long run, however, the invention served to reinforce bureaucratic institutions. The civil service examination system in particular, with its thousands of answer books, would have been inconceivable without it.

Competitive public examinations for official service began in the Sui (622 A.D.) and were given at intervals of one to three years. This system replaced an earlier method of appointment which relied heavily on personal recommendation. It seems to have been inspired by a desire to strengthen the central government of the monarch at the expense of cliques and wealthy families—a desire similar to that which prompted the Meiji government in 19th-century Japan to counter the power of the cliques from Satsuma and Chōshū domains by drawing talent into the bureaucracy through the law department of Tokyo Imperial University and the Higher Civil Service exams. In any event, the consequences were unmistakable. The examination system linked scholarship directly to national job opportunities; as it demonstrated the

[1] In general, the dissemination of inventions is not dependent solely on technological factors. In the absence of a social demand, inventions are not developed and on occasion even forgotten. Anthropologists report that primitive peoples have often expressed no interest in a modern invention with which they happened to have come in contact and that the experience quickly disappears from their memories.

capacity to provide authoritative public standards for judging among prospective candidates, respect for learning was intensified and broadened among the elite. This was the beginning of a meritocracy, the genesis of a system in which the examiner is always more important than the scholar.

The civil service examination system was obviously a boon to academic endeavor. Yet the more demanding the examination, the more formalistic it became, so that scholarship inevitably came to be a matter of memorization and recitation. Interpretations of the classics grew fixed, and the importance of being able to recite encyclopedic bodies of information created an interest in the compilation of indexes and reference guides, books that consolidated this information and made it more readily accessible (Miyazaki, 1946).

The examinations themselves consisted of both oral and written sections, the former being chiefly a test of the examinee's ability to memorize. One of the favorite Chinese methods of doing this was to select a line of classical text, cover the surrounding passage plus three characters in the line itself, and ask the examinee to recover the missing characters. In the written portion of the exams, scholarly essays and discussions of policy left some room for originality.

The explicit purpose of the system lay neither in education nor in the promotion of learning but in choosing a limited number of men for official posts from among a large pool of candidates seeking to improve their lot in the world by rising through the ranks of public service. In other words, it was not to train officials, but to select them. Questions were standardized in the interest of fairness so that the knowledge expected of examinees had a high degree of uniformity. Thus, while the Chinese examination system played a positive role in the rise of a literati class and the development of the bureaucratic generalist, it contributed nothing to academic specialization or the discovery and growth of new knowledge. On the contrary, it operated to imprison scholarly endeavor within certain standard limits.

Of course, inasmuch as academic inquiry is ultimately an individual activity, it cannot meaningfully be evaluated through written exams designed for large numbers of persons. The material from which examination questions are drawn is always at least one step behind the cutting edge of scholarship. A new body of knowledge or mode of learning can serve as a source of general examination questions only after its generative period is past and it has acquired a certain stability. Thus the *avant-garde* Neo-Confucianism of the Sung was not incorporated into the civil service examinations until the Ming, by which time it had become a bulwark against new philosophic trends.

If institutionalization is the final stage in the social history of a paradigm, as I suggested in the previous chapter, institutional structures are normally the products rather than the progenitors of unprecedented developments in scholarship. The founding of schools and the creation of an educational system does not necessarily lead to the growth of learning. As an institution, the school incorporates and maintains knowledge that has been developed from a given paradigm. A vessel which shields the lamps of learning from the winds of external pressure, it also serves to check and control explosive internal developments. Thus one can say that new modes of thought and fields of academic inquiry have very little to do in their nascent stages with the institution and collection of buildings we call the university. (I shall elaborate this argument in the following chapter.)

The internal control function of institutions is particularly striking in the Chinese case. The examination system in which Chinese scholarship found its institutional home was for centuries the most rational system anywhere in the world of recruiting an administrative elite, and it provided the greatest equality of opportunity for aspiring officeholders. The system was so well constructed that even members of the educated elite who occasionally harbored unorthodox ideas or would have devoted their efforts to regionalist competitors of the government were channeled into the path to worldly success that it represented and found their intellectual energies absorbed within the system. It is one of the ironies of history that the institution proved to be so well made, so stable, and so long-lasting that it successfully blunted the will to reform and revolution. Dynasties came and went, but radical change in society proved impossible as long as the system of recruiting the bureaucracy through literary examinations continued. When the institutional framework within which scholarship was carried on hardened, learning itself ossified, and an academic tradition that favored documents and memorization over disputation became a dominant and permanent feature of Chinese society.

It can be argued that the biggest reason for the stagnation of Chinese science was its geographical isolation from Islam, India, and Europe and its consequent insulation from the challenges of a heterogeneous culture. Yet there was considerable contact with the West during both the T'ang and Yuan periods. The point then is not so much that China was wholly without opportunities for this kind of encounter as that the challenge was either rejected or ignored. The standard explanation for this response is the Chinese belief in the superiority of their own civilization. This argument is too crude and racist to be fully persuasive, but our discussion of paper and the bureaucratic system may point the way toward a more adequate explanation. For the Chinese of this period learning was something that was written on paper, something printed. Though they came in contact with many strange artifacts from the West, they encountered nothing that fit this description, nothing that corresponded to their understanding of what scholarly activity was about. Could there be scholarly activity where there was no paper? And could a higher civilization exist where there was no scholarly tradition to convey and preserve it?

The Pre-Paper Culture

What kind of relationship existed between paper and learning in the West? The first known way station in paper's westward journey was the Central Asian city of Samarkand, where paper was first manufactured in 751. Carried by the Islamic world, it entered Europe through Spain in the 12th century and spread slowly across the continent. This suggests at least a *prima facie* connection between the dissemination of paper and the revival of European learning which was taking place at this time under the impact of Islamic science. Paper may even have played a major role in this Renaissance of the 12th century. Whether the revival of Greek learning stimulated the manufacture of paper by increasing the demand for large quantities of copy materials or whether the advent of paper simply paved the way for the revival of learning has not as yet been ascertained, but they clearly traveled side by side.

Parchment had been the West's chief writing material ever since the days of Rome, though the ancients had used clay tablets and papyrus.[2] Made from sheepskin, parchment was more durable than papyrus or paper, but it was also much more costly. One can get some idea of the cost of a parchment book from the fact that the manufacture of a 200-page volume required the skins of 25 sheep. This no doubt explains why parchment appears to have been used initially only for important religious and legal documents. It may also help us understand why the idea of printing does not seem to have seriously occurred to anyone until paper came into use. The cost of labor to copy manuscripts was so much less than the cost of parchment that it was not even considered significant (Clopham, 1957). Thus professional and part-time scribes were as essential to scholarly activity in the medieval university as printing and experimental apparatus are to contemporary research.

Yet the introduction of paper and printing did not mean that books were immediately available to the average student and before this came to pass, Islamic and medieval scholarship had already acquired a clearly defined institutional form. These circumstances are evident in the maturing pattern of higher education in the West, a pattern that antedates the use of paper and printing. (Even today some universities in the Western world use sheepskin for their diplomas.)

The Islamic Madrassa

We are accustomed to thinking of the system of higher education that exists today as having its roots in the universities of medieval Europe (Rashdall, 1936). But Islamic scholars argue plausibly that its origins should be placed even earlier, in the Islamic *madrassa* (Sayili, 1948). They point out that it was during the 12th and 13th centuries, when the influence of Arabic scholarship was strong, that universities were first built across Europe, and they

[2] According to technology historian Lynn White, Jr., the shift from papyrus to parchment came about because of a sudden drop in the production of papyrus due to blight.

note certain similarities in architectural style, in layout, and in the fact that both centered around a system of conferring degrees.

Typically situated near a mosque, the traditional *madrassa* continues to exist today alongside modern universities of the European and American types, transmitting a culture that reached the peak of its brilliance during the 9th to 11th centuries. Laid out very much like a Cambridge college, the *madrassa* houses students in rooms clustered around central gardens while the center is reserved for a place of worship.[3] It is in this central area that students can be seen sitting around their teachers, engaged in seminars called *halqa*. These schools first emerged in the latter half of the 11th century as institutions for higher studies in theology and law; their development came at the very end of the creative period of Islamic science (9th–11th centuries), confirming our suggestion that institutionalization normally follows outbursts of new learning.

Financially, the *madrassa* was the beneficiary of income from property trusts known as *wakf* set up by the wealthy pious. This income supported teachers and other employees and made it possible for the student to receive his education, along with board and room, free of charge. Usually he was also given a small stipend. Each *madrassa* had its own library, and some had hospitals and bath houses as well. Others, established at the palace or within the King's quarters, were luxuriously appointed. What was taught had a religious flavor, but the *madrassa* was not isolated like a monastery, and topics were discussed that we now associate with a secular, liberal education. The study of philosophy and science was not an explicit aim as it was in the European universities, yet it was Aristotelian logic that set the fundamental tone of the *madrassa*'s seminars.

According to Al-Farabi (ca. 870–952), ascendant Islam divided knowledge into five categories: (1) the study of language, including composition, grammar, pronunciation and speech, and poetry; (2) logic (Aristotelian); (3) preparatory studies, including the Platonic Quadrivium and some Hellenistic sciences—arithmetic, geometry, optics, astronomy, harmonics, the study of weight, tool and utensil design; (4) physics (the study of nature)

[3] I have seen early Buddhist architecture in Afghanistan that resembles the *madrassa*.

and metaphysics; and (5) social studies—law and rhetoric. This scheme indicates the degree to which the rhetorical-logical tradition of the Greeks remained alive in Islamic culture (Nasr, 1968). Within this tradition, anti-Aristotelian discussions went on and such heterodox notions as heliocentrism were entertained. By contrast, the learning carried on in the schools and reflected in the civil service examinations of T'ang China shows little evidence of the logical impulse. Classical exegesis, chiefly of the Five Classics of the Confucian canon, was the main scholarly activity. The other special examination fields (medicine, arithmetic, calligraphy, law, etc.) formed a list with little integration and design save that dictated by the government's need for bureaucratic specialists (Taga, 1953).

In the 9th century, Islamic society experienced a scientific awakening that has been explained as a response to the inability of Muslims to hold their own in debates with the Christians and Jews they encountered on the streets of places like Damascus and Baghdad. The study of Greek philosophy and science was, in other words, prompted by the need for logical weapons (Nasr, 1968: 70). By the early 11th century, however, this creative period in Islamic science was drawing to a close. Construction of *madrassa* peaked at the end of the 11th century, but by this time science was neither considered important nor included in the curriculum. Islamic learning settled down to a fixed pattern and started down the road to ossification. Institutional growth did continue for a time, and hospital and observatory facilities were expanded (Sayili, 1960). Under the patronage of rulers with astrological interests, large observatories were built at Maraga and Samarkand, contributing to the refinement of both observation and theory. Thus even in the ebb tide of creativity, a well-organized institutional system ensured the ongoing accumulation and refinement of data, the continued growth of normal science. In the 14th and 15th centuries, however, even institutional growth came to a standstill and institutions began to lose their vitality, though they continued to function as preservers of traditional scholarship.

The Universities of Medieval Europe

The medieval European university differed more from its Islamic forerunner in certain of its formal features than in the substance of the lectures and discussions that went on within its walls. At the heart of these differences lay distinctive systems of conferring degrees. Islamic society granted licenses to teach called *izaya*, whose holders probably shared some of the attitudes of the professional guilds of Europe. The economic conflicts that broke out in Europe between students and masters and sometimes spread to the general citizenry were not in evidence, since both teachers and taught received their support from the *wakf*. The Islamic system of degrees also lacked the elaborate legal distinctions that emerged in Europe.

The *izaya* degree was ordinarily awarded by individual teachers to students they considered sufficiently accomplished to teach a specific classical text, so that the teacher with whom one had studied was more important than the *madrassa* attended. Students moved from place to place in search of teachers and often equipped themselves with licenses from several (Sayili, 1948: 46). In the European university the *licentia docendi* was issued in the name of the institution or its patron to students who had mastered an entire course of study, and it entitled them to membership in the teachers' guild. The legal rights and privileges it conferred were clearly stated. Eventually it enjoyed a wider function: after the universities effectively institutionalized scholarship, university degrees became prerequisite requirements for entry to the clerical office and other official posts.

The "early Renaissance" of the 12th century manifested itself differently in different places. At Bologna, one of Rashall's two archetypical medieval universities, it was a revival of the study of Roman law, while at the other, Paris, it was an outburst of dialectical speculation in theology. Both these developments may well have been directly influenced by the Islamic *madrassa* where theology and law were the chief studies. In any event the whole movement was grounded in the study of logic.

> Logic was the one treasure salvaged from the intellectual ruins of the past and students were encouraged to make it their own. . . . Without knowing it, they were led step by step from logic to the study of nature, from natural science to metaphysics, from metaphysics to theology. . . . (And) before anyone was quite aware of it, a universally applicable theory had been constructed (Rashdall, 1936).

Logic knows neither heterodoxy nor orthodoxy. Neither does it require a sensibility cultivated and refined in a particular cultural climate, like poetry. It possesses a universality that makes it intelligible even to "barbarians"—though medieval logic might also have led on occasion to barbarous results itself. It has been said that the primacy of logic and debate in medieval intellectual life adversely affected standards of classical literary style (Abelson, 1906). But the success of logic during this period was not the result of any revolutionary change within the discipline itself. Logic continued to mean Aristotelian logic. What happened was that, beginning with the intellectual sensation created by Aristotelian learning in the 13th century, the logical habit put down roots. It was in fact by virtue of its powerful logic that Aristotelian learning, never wholly dominant in Hellenistic circles, survived shifts in the center of learning and emerged as the chief pillar of the Western academic tradition. And as this Aristotelian learning entered the universities, logic acquired a commanding position in the Trivium, while the Platonic Quadrivium that had heretofore been preserved in the monasteries went into eclipse.

Logic was an important subject in its own right, but it was more than that. It found its way into law, medicine, philosophy, and theology, making its presence felt in all fields of intellectual endeavor and shaping the character of the medieval mind (Haskins, 1923). If we understand the growth of modern science as a process in which the methods and approach of physics spread to the several sciences, medieval scholarship might be described as a process in which logic came to permeate all fields of learning. It became, in other words, a model science.

In terms of method, oral instruction inevitably predominated, for printed books were not available and copied manuscripts expensive. Students made their own books by taking down the dictation of their masters. The medieval university was particularly enamored of the dialectic, often taught by dividing students into affirmative and negative groups for discussion of particular topics. Examinations were also oral and often required that a proposition be challenged or defended. The examiner functioned as a judge, making sure that each student had adequate opportunity to present his argument and seeking to keep the discussion from wandering off on a tangent. This disputative style was an integral part of students' daily lives as they argued, in the words of one writer, "before, during and after meals, at all times and places whether public or private" (Tillyard, 1913). Men like Johannes Buridan, Rector of the University of Paris, engaged in the 14th-century debate over the notion that the earth was in motion "purely for the sake of a good argument." And when Nicole d'Oresme (d. 1382) sided with the supporters of the theory that the earth turned on its own axis, it was not, he made it clear, because he believed it to be true (Beaujouan, 1963). An element of the medieval tradition remains in existence today wherever degree candidates must successfully defend their theses against the challenges of several members of the teaching guild before gaining admission to their fellowship.

In placing a premium on the capacity to memorize, the Chinese examination system forced students to go over the classics so early and often that they tended to become absolutes, and scholars lost both the power and the will to criticize them. The system also made possible government schemes of thought control that focused on regulating the content of the examination. The rhetorical style, in contrast, invited criticism of authority. Even Islamic science, once described simply as a deep-freeze for Greek learning, appears upon closer examination to have produced its share of anti-Aristotelian criticism and its own debates over heliocentrism. Nor was the recurring criticism of the Aristotelian paradigm that emerged from within the Scholastic tradition without cause—though it is true that this criticism remained largely within the categories of Aristotelian logic and did not overturn Aristotelian learning itself.

Yet the disputative approach is not without its limitations as a scholarly method. Its immediate goal is the skillful use of logic to persuade others to accept theories and interpretations. It encourages men to address themselves more to their opponents than to "nature" or "things." The critical spirit may well flourish under these circumstances, but it easily runs unchecked for want of reliable criteria of truth. It was the genius of early modern science to make nature the judge of veracity, and it was by stressing evidence over argument that it finally triumphed over the Scholastic logicians.

If the Schoolmen were not yet natural scientists in the modern sense, they did approach nature with the eyes of lawyers. Many scholars both within and without the Western world have sensed an affinity between the juridical notion of natural law and the laws of nature and have sought to explain it in terms of a common origin. Joseph Needham attributes both to the special features of monotheistic religion. The laws of nature are those laws to which a transcendent personal Deity subjects the natural world, he argues, comparing Him in this respect to an absolute monarch whose word is law among his people (Needham, 1956). It is true that the Chinese lacked the type of transcendent, personal Deity that might have aided the growth of a full-blown sense of nature obedient to strict laws. But they believed in and searched out mathematical regularities in sky and earth nonetheless. Is the link between natural law and the laws of nature not better understood in the context of the whole argumentative tradition of the West—in terms, that is, of a tradition of classical rhetoric, Roman law, and Scholastic logic? The presence of monotheism has not always led to a search for what scientists would call laws of nature, while science has flourished in places such as Greece where the gods were many and there was no concept of law inherent in nature.

In gaining the assent of any group of scholars to a common paradigm, nothing is so convincing as natural phenomena. If the social sciences still remain argumentative disciplines and continue to resist transformation into normal sciences, it is because the phenomena from which data are derived in these fields do not express themselves in unequivocal form. If one had to say in a word how the science of the 14th century as practiced by the Schoolmen differed from the science of the revolutionary 17th century, one might observe that medieval science was preoccupied with philosophical questions of method applicable to any imaginable case, and had relatively little interest in the much smaller set of phenomena that could be observed in nature (Crombie, 1961). Scholastic science had grown out of discussion and debate. This fundamental characteristic was to give Scholasticism a preliminary role in the unfolding drama of modern science, but finally kept it from starring in the crucial sequence.

Yet from the perspective of a specialist in the history of East Asian science like Mikami Yoshio, who finds in the absence of this kind of logic the great weakness of Oriental mathematics and science, portraits of the scientific revolution that stress the role of mathematics and the experimental method over that of logical rigor are not wholly persuasive (Mikami, 1926). Exclusive concern with formal logic may have been one of the evils of the Aristotelian tradition of which 17th-century scientists were determined to rid themselves, but of all the preconditions of the rise of modern science, it is the logical character of the West that attracts the attention of those who are able to look at European intellectual history from the outside.

The Movement of Scholars

The early dissemination of paper and printing in China was not necessarily an aid to creative scholarly activity. The function of any given technological invention differs according to the time and place in which it appears. In China printing had become common by Southern Sung times (1127–1280), so that even someone in a rural area could buy one of the printed reference books designed for examination candidates and study it by himself (Li Kuo-tiao, 1956). But in the medieval West, where the lack of printing techniques made conversation and scarce manuscripts indispensable for study, scholars and students travelled from town to town in search of both. One can readily imagine the degree to which these journeys quickened the mind and accelerated the exchange of information.

The *Sankin Kotai* system of Tokugawa Japan and Muslim pilgrimages to Mecca helped to broaden men's knowledge of the world around them, to bring them fresh intellectual challenges and to change their way of thinking. Even today study abroad often leads to a change in one's basic approach to his subject. What is true of the effects of new encounters upon individuals seems also to have been true of culture areas. Isolated at the eastern end of the Eurasian land mass, the Chinese cultural sphere did not receive major challenges from the outside world and never went beyond the accumulation of routine knowledge, while Western learning experienced a variety of revolutions as the center of the tradition shifted, periodically bringing it face to face with major new challenges. Had the center of the Western academic tradition remained in Greece, it probably would have come to look more like the Chinese and would have proved incapable of generating modern science.

Generalization

Encounters with new things and different values tempt the will to grasp experience in some sort of integrated fashion; this moves scholarly endeavor toward generalization. Traditional Chinese learning, as has often been observed, seems to have lacked this drive toward generalization. The Chinese observed and they put their observations down on paper. But in doing so under a dominant academic style that encouraged and rewarded exhaustive documentation, they became increasingly unable to integrate their information and use it for their own purposes. In the end, that most primitive of all normal science methods, the keeping of records, ruled supreme.

In the most rudimentary sense, generalization is simply the attempt to catch a wide range of phenomena in one net. As a scholarly act, however, it frequently involves an awareness of other intellectual positions and an attempt to integrate them dialectically with one's own. Nevertheless, it does not necessarily begin by surveying the scholarly landscape with equanimity, giving the same weight to all phenomena that come within its purview. Typically the scholar takes his departure from his own field, his special territory, his own particular research. Newton, for example, set forth the law of universal gravitation from his base in planetary motion in an attempt to extend his insights to previously inexplicable phenomena. Operating from different turf, Cartesians (with their interest in collisions and quantitative movement) and the followers of Leibnitz (from their inquiries into falling bodies and *vis viva*) moved toward the construction of their own kind of dynamics.

Successful generalization thus involves being able to explain phenomena in fields outside one's own in a convincing manner. When this means an attempt to gain adherents for a new paradigm by using it to explain phenomena that have heretofore been dealt with under another, the task is one of development and enrichment. But the application of generalizations from a special field of research to areas of scholarship that have yet to be developed serves to set new problems for normal science and stimulates the growth of a new scholarly tradition. To the "empirical" practitioners of documentary learning, however, neither of these tasks is familiar. It is rather the addition of a commentary, exegesis, or historical document to the mountain of accumulated knowledge that is regarded as meritorious. This act does have value in itself. But that value is wholly intrinsic. It resists generalization and hence transfer to other problems and situations.

Classifications of Knowledge

The geographical flow of learning through different cultural areas also seems to have had its consequences for classifications of knowledge. In China, the library classification system that had been developed by Liu Hsin at the end of the 1st century B.C. continued to be used without basic modifications down through the Ming and Ch'ing periods. In the West, however, Islamic scholars undertook several revisions of the Aristotelian system. The early work of men like Al-Farabi culminated in the 14th-century classification of Ibn Khaldun, which closely followed changes in the *madrassa* curriculum. It incorporated several new fields of study that had developed in the Islamic world—Koran studies, religious law, theology, and Arabic linguistics. In the medieval European university the influx of Aristotelian learning into a curriculum built around early Christian theology (particularly in the theological faculty) also led to a reorganization of knowledge. In short, when the center of scholarly activity shifted to another cultural area, patterns of life and thought collided, stimulating the reorganization of knowledge and making the problem of classification an issue.

Once fixed, classifications of knowledge help to preserve existing knowledge, but they impede the emergence of new disciplines. In the Chinese cultural sphere, where the classification system was not reorganized, old knowledge was well preserved, but those fields of inquiry that were not included in the initial scheme never became fully legitimate. The Chinese language does not lack grammar and logic, but the fact that these two subjects did not develop into independent disciplines in the world of Chinese scholarship was to make the learning of the East significantly different from that of the West (Graham, 1973: 64).

The Consequences of Printing

The revolutionary influences which flowed from the dissemination of printing hardly need lengthy recitation here. It may be noted, however, that printing resulted in a wholesale transformation in the scale of values by which scholars were judged. In an earlier period a man could become a respected scholar in the West by talking, lecturing, and cultivating his persuasive powers. With the advent of printing, however, the scholar was motivated to think about getting his work into written form—a trend that has continued until today, when "publish or perish" is one of the ground rules of academic competition. This was the beginning of the end for the Platonic style of argumentative learning, which regarded writing as deleterious to the human spirit, but it was also a step toward broadened participation and equality of opportunity in the academic world.

Printing not only makes possible a wider readership: it requires it. Moreover, writing for publication means addressing a large, unspecified audience. From a traditionalist point of view it has been said that printing began as a vulgar medium which catered to the masses—a criticism later made of early radio, motion pictures, and television. When printing appeared in China in the 10th century, it was first widely used for items of everyday popular interest—popular tracts, Buddhist prayer books, and almanacs. The orthodox Confucian classics did not find their way into print until about a century later (Li, 1962). Nevertheless the tremendous impact of the new wave of Neo-Confucian scholarship on the intellectual world owes much to the printing medium.

The year 1543 is generally remembered in the history of science as a milestone on the way to modern science, for it marked the publication of both Copernicus' *De Revolutionibus* and the *Fabrica* of Andreas Vesalius. Copernicus' advocacy of heliocentrism gave his work its notoriety, but the notion that the earth revolves around the sun had had proponents in Greece, India, and the Islamic world as well. The particular significance of Copernicus' book lay in the astronomical system that he built upon the heliocentric hypothesis, but no less in the fact that such a grand design had now been put into print. As for Vesalius' account of human anatomy, even historians of science have difficulty finding anything really revolutionary about it. There is no other way to explain the impression he made on the general reader except to say that getting his remarkable work into print when he did made him first in the field.

The Italian universities, and Bologna in particular, admitted surgeons to full membership in the guild of masters, breaking away from the customary contempt for those engaged in practical work that lingered on at Paris and most of the other universities in Europe. Yet Italian doctors made few contributions to progress in biology, or even in fields such as anatomy and botany where their work would seemingly have given them an edge, for they lacked appropriate means of transmitting their observations (Overton, 1957). Manuscripts were copied by large numbers of students at once, following a pattern developed in the medieval monastery. While the text was read aloud, each student made one copy. (Even in this age of copy machines, we are still following the medieval system of lectures and note-taking!)

Had the texts been in the complex ideographic notation of Chinese, this "mass production" method might not have worked so well, but it was particularly suited to the phonetic systems of the West. Still, with diagrams there were complications. The parchment and the heavy paper of the West could not be used for tracing, and in an age when there was a dearth of other reproduction techniques, the omissions and personal quirks of the scribe began to take their toll on fidelity. Copies began to diverge from the original, and the discrepancies grew larger as copies were made from copies. The dissemination of scholarly work in anatomy and biology proved, in fact, almost impossible before the block prints and movable type of Vesalius' time. Unpublished, the anatomical drawings of Leonardo da Vinci did not, for all their brilliance, attract wide attention. Need more be said about the importance of printed books to the scientific revolution in academic style?

A Written Culture and a Spoken Culture

Where then do these observations concerning the material

media of academic communication and the substance of academic culture lead us? Can it be maintained that it was the early invention of paper which impeded the development of logical thinking in China? The proposition that whole academic traditions were essentially determined by the communication technology available to them at crucial stages in their development is not historically tenable. The Chinese bureaucratic institution was in existence before the advent of paper, and, as I have observed in the preceding chapter, the rhetorical tradition of the West and the record-keeping tradition of China had already taken clear form prior to the diffusion of paper and printing. Yet there can be no denying that the available technologies of communication reinforced, exacerbated, and enlarged the differences between the two traditions.

Several years ago Japan was astir with the McLuhanist notion that television and other new communication media were exercising a decisive influence on patterns of human thought and action. Marshall McLuhan based many of his observations on a comparison of the oral culture of a pre-literate African people with the demonstrative culture of the West which he found to depend upon the technology associated with a phonetic alphabet (McLuhan, 1968). As far as academic society is concerned, however, the West seems much more verbally oriented than China. To this one might add that Chinese ideographs are more visual than the phonetic alphabet and more easily fixed in the mind. Japanese, who in many ways are peculiarly qualified to judge since they use both Chinese ideographs and phonetic *kana*, and are familiar with the Latin alphabet as well, often feel unsure of themselves when forced to rely on hearing alone for the name of a person or a thing. Identifying the spoken word with its Chinese characters is necessary at such times.

Ideographs are also of particular utility in retrieving written materials. This proved to be a major advantage in ordering and classifying the documents of an annotative and exegetical tradition, and undoubtedly contributed to the important role played by "library series" and "indexes" in the growth of Chinese scholarship. For Japanese, *kana* mixed with Chinese characters can be read more quicky than *kana* alone, and the presence of characters makes it easier to decide what the passage is all about even if one does not read them all correctly. Chinese characters are also more useful than *kana* for title pages, tables of contents, indexes, and other places where one wants to know at a glance what is there. One might say that the phonetic alphabets of the West represent auditory languages that are well suited to debate and conversation-centered argumentative scholarship, while Chinese ideographs are a visual language most useful for recording and retrieving documentary material.

In the progress of the human race, writing is clearly a more advanced act than speaking. Yet before coming to any conclusions about their relative superiority we must examine the ways in which these two means of communication have functioned in particular ages and civilizations. The challenge of the written word is at best an indirect one. A book is written, printed, and comes to the attention of the author's intellectual peers and adversaries. Only after it has been read does any criticism get back to the writer. In the interim, his perspective on the issues and problems he addressed in the book may have changed or grown fuzzy. Certainly some time will be required before he is able to take account of criticism and reformulate his argument. For the reader as well as the writer, the culture of the written word is cast in the past (or perhaps even the past perfect) tense. Oral communication, on the other hand, is cast in the present progressive tense. Responses are received on the spot. Generically impermanent, the spoken word vanishes into thin air, but this gives the speaker the freedom to venture whatever new thoughts come to him without being unduly afraid of saying something wrong. Thus it is in the midst of oral exchanges, of dialogues and debates, that academic life ordinarily is most filled with tension and excitement.

Today the scientific community formally recognizes new achievements only after they have been published. The printed page becomes a kind of record of the achievement which, if properly done, presents research results in a manner that makes them readily accessible to others, and narrates the process by which the results were obtained in a persuasive (if idealized) style. But is this the sole means by which scholarly traditions can be conveyed? And how effective is it as a method of communication?

These are important questions which those interested in the

history of science would do well to ponder. Can scholarly traditions be adequately transmitted through the printed word alone? Or is direct contact and discussion indispensable? To bring the matter closer to home, can science still afford to rely chiefly on the printed word?

In the West, the introduction of paper and printing led to the death of the Greco-Roman rhetorical tradition. What was to replace it as a stimulant to academic vitality? In the following chapter I will discuss the contributions of academic societies and scholarly journals in this regard. It may be noted in passing, however, that the introduction of paper and printing in the West did not result in any sudden changes in the educational program of the medieval university. Written examinations were apparently not held until the 18th century, the one given by Richard Bentley in 1702 being the first on record (Gotō, 1933). As for the Chinese examination system, the French Jesuits who visited China in the early Ch'ing praised it highly (Teng, 1943). Certain Englishmen also encouraged its adoption. Although a written examination was not employed in the selection of European government officials before the French revolutionary government adopted the practice in 1791, by the 19th century written tests were widely used in the recruitment of an administrative bureaucracy in the English colonial empire and elsewhere. If voices could now be heard deploring the adoption of a system identified with a backward country like China, they were simply testifying to the real influence of the Chinese examination system on the West.

The things that practically never reach print (the reasons one takes up certain questions; inside stories about successes and failures in research; and so on) would seem to be extremely important to the maintenance of a creative academic tradition. Objectively, one can say that intellectual contact through books and printed materials provides an opportunity for dispassionate reflection. Yet the flesh and blood of a fashioner and maker of learning is not readily communicated on a piece of paper.

But what are the academic conditions under which scholarship ferments and becomes creative? The answer, pure and simple, is the proximity of companions with whom one engages in serious academic discussions. Access to a wealth of documents is a necessary but seldom sufficient condition. Printing, we are told, heightened the distinction between writer and reader. In television as well, a thick wall exists between the performer and the audience. Thus the division between teacher and taught that was part of the medieval lecture system has been institutionally preserved and technologically reinforced. Does creative dialogue survive only in the seminar system of the modern university?

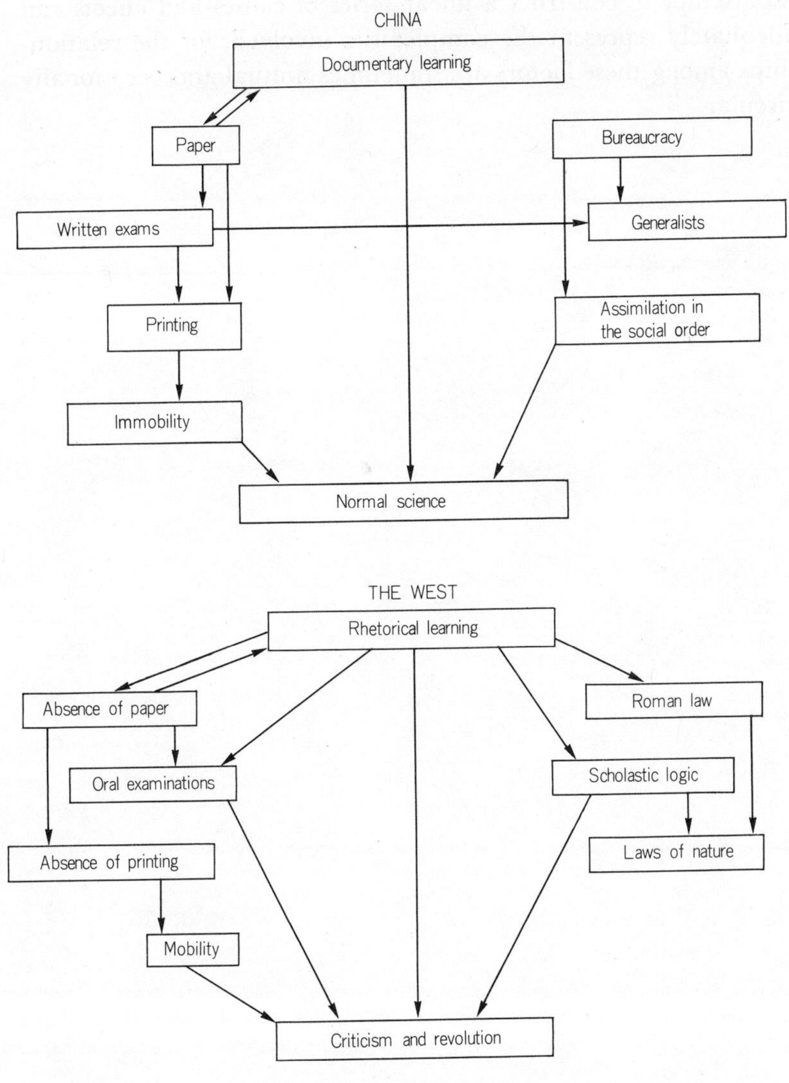

Figure 3–1

A Diagrammatic Summary

Thus far I have set out several factors that seem to contribute to an explanation of the difference between the academic traditions of East and West. To facilitate an understanding of the cause-and-effect relationships involved, I have put all these factors together in Figure 3–1. As the shape of the diagram and the arrows suggest, no attempt to construct a linear series of causes and effects can adequately represent the complexities involved, for the relationships among these factors are sometimes mutual and occasionally circular.

Chapter 4

Scholarly Journals, Learned Societies, and the Rise of Modern Science

At the end of the Middle Ages, and even into the Renaissance, there were no decisive differences between the academic traditions of East and West. Although Joseph Needham's monumental study *Science and Civilisation in China* (called by one critic the greatest work of the 20th century) was initially inspired by the question of why Chinese science stagnated and was hopelessly outdistanced by the West in the early modern period, the author has recently confessed that a second question now looms larger. After many years of research he has come to ask why, during the long period between about 250 B.C. and 1450 A.D., the East was superior to the West. Many Western scholars—and their Asian counterparts—have tended to see this observation as a product of Needham's love affair with China, and have dismissed the view of history it implies accordingly. Yet, as we have seen, until the European Renaissance, the Chinese academic tradition was roughly analogous to—and could be meaningfully compared with—that of the West. It was in many ways a different kind of tradition, but its distinctive features seemed no more significant than the variations in clothing and eating habits one expects to find in different climes and places.

By the time modern science came into existence in the 17th century, however, the gap between China and the West had become a chasm. In seeking to explain it, however, it would seem appropriate to ask, not why China stagnated, but why the Scientific Revolution occurred in the West. To illustrate the problem metaphorically: one can analyze the causes of a fire, but there is an obvious fallacy in trying to specify the reasons why, at any particular time and place, no fire occurred. It is, in other words, not the fate of the Chinese tradition but the Scientific

Revolution that is the intellectually extraordinary phenomenon and which is, therefore, more likely to reward our search for causes.

The Useful Knowledge of Things

During the 1930s, historians of science argued about whether Scholastics or artisans had made a greater contribution to the rise of modern science. Academic orthodoxy emphasized the importance of Scholasticism while Marxists stressed the role of the artisan, but by the time the controversy had died down, it was apparent that modern science had come into being not as the offspring of one or the other of these traditions but precisely at the point where the two had come into contact. Individuals could still attach different relative values to the two streams, but their sentiments could not be conclusively quantified and compared. Even so, the Marxists were somewhat hampered by the nature of their nominees. Craftsmen are not as verbose as scholars. In general they are not particularly given either to reading or writing. Thus the Marxists found it difficult to document their position and were forced to rest their arguments on circumstantial evidence drawn from social history.

But another group of men appeared in Europe at this juncture, who were neither silent craftsmen nor Schoolmen wholly absorbed in the world of words. They were educated men who participated in the speculative and rhetorical tradition, yet took an interest in the traditions and practical work of craftsmen. Several factors may have prompted them to observe the workshop and record what they saw, but they brought to this exploration an intellectual appetite. These intellectuals made light of the endless, vacuous discussions they found in books. They engaged in and encouraged the pursuit of useful knowledge to be acquired directly through the study of the phenomenal world.

Had it not been for these scholars, the traditions of artisan and Scholastic might have run their separate courses indefinitely and modern science never have come into being. But these students of things were more than intermediaries. They brought the two traditions together in a way that both set them aside and transmuted them into a higher synthesis. Francis Bacon, an ardent proponent of this new orientation, once observed:

> Even though what commonly passes for logic has been correctly applied to the technology that is rooted in social life, language and popular opinion, it is incapable of arriving at an understanding of the intricacies of nature, and in reaching for that which it can not grasp, it has done more to harden error than to open the way to truth, aggravating, so to speak, its entrenchment.

His own logic was to be different:

> In ordinary logic, opponents attempt to corner one another through debate, but in this [Bacon's] logic it is nature, through her operations, which strikes and binds us (Preface to *The Great Instauration*, 1620).

Though Bacon's formulations were often peculiarly his own, one can find the same concerns in Galileo, in Boyle and Huyghens, and finally in Newton. These men were not wholly immersed in the artisan's world of half-secret traditions, but they took a strong interest in his technology and sought to record the fragmentary knowledge that was the fruit of his experience. The artisan's interest in the miscellaneous operations he performed seldom extended beyond the finished product of his craft, but as men with broader vision undertook to put his perceptions in writing, an interest in classifying and ordering them was born.

Yet these kinds of records were not necessarily new in Western civilization. The encyclopedic writings of the Roman general Pliny witness to that. One can also find them in China. Indeed, useful knowledge, experientially acquired through observation and manipulation of things, is part of both the Chinese and Japanese traditions. The editor of the *T'ien Kung Kai Wu*, the Ming local official and collector of technological information Sung Ying-hsing, and the Japanese astronomer and calendar maker Nishikawa Joken were all interested in the knowledge of things. In fact, Chinese and Japanese writers were often closer to both nature and the workshop than the Schoolmen of the West. The Chinese scholar-official had the normal administrator's interest in technological problems, though his vantage point was always that of a manager rather than a producer.

But modern science is not simply an extension of this kind of orientation. There are those who would take the progress of the positivist and empiricist spirit as the sole criterion by which to measure the march of mankind toward a truly "scientific" civilization.[1] In this enlightenment view of history, science and superstition are treated as antithetical concepts whose confrontation is the history of human progress. In the Dark Ages, the surface of the earth was covered with the black stones of superstition, but these will be replaced one by one with the pure white stones of science until mankind arrives at an ideal, gleaming world.

There is also a strain of thought which holds that it is only with modern science that men have acquired the criteria by which to distinguish between what is and what is not superstition. Yet aversion to superstition is at least as old as Confucius' determination "not to speak of the supernatural." The Roman Seneca also refused to accept the preternatural and mysterious as proper subjects for scholarly inquiry. But modern science is not a child of this secularizing common sense. Indeed the anti-commonsense, occultist attitudes of a Kepler are more typical of the kind of sensibility from which it emerged. For if the notion that the earth moves violates common sense, the idea that the earth and the moon pull against each other dwarfs the circumscribed concerns and perspectives of ordinary life.

A simple positivism existed as long ago as ancient Egypt. Early Egyptian medical papyri stick extremely close to experience, more so than later manuscripts which tend to be more speculative. Nor should astrology and alchemy simply be dismissed as pre-empirical, consisting as they do of accumulations of positive, recorded knowledge subsequently reordered within a superimposed framework of notions related to the prognostication of fate or inspired by the possibility of eternal youth. One who wishes to call them superstition must also admit that in these cases positive data proved

[1] Science historians consider the peculiar mathematical and mechanistic view of nature embodied in the work of Galieo and Newton as the defining characteristic of modern science, but general historians, particularly Japanese historians, often tend to identify modern science with Bacon's "practical knowledge," and to suggest that it is something to be attained simply through the continued accumulation of positive facts. For an example of this point of view see Sugimoto (1962). One of the problems with this position is that it has no way of dealing with the revolutionary character of modern science.

to be a seedbed for something quite unscientific. Yin-yang and Five Elements theory and the Greek notion of the Four Elements also represent attempts to systematize positive data, so that they should not be condemned outright as retreats into superstition.

In short, even if factuality is to be considered a preliminary indicator of the degree to which a given theory or tradition approximates modern science, the claim itself is not meaningful unless one first specifies the context. Modern science may be more factual than Aristotelian mechanics, but to place factuality first on a list of the distinctive features of modern science is, I think, to miss the point. Before it could become modern science, the useful knowledge of things had to pass a qualifying test. In the standards by which it was measured in this test we shall find the distinctive features of modern science.

The Uniqueness of Modern Science

As we have indicated, the 17th-century Europeans whose quest for tangible truths led them back to the study of things themselves were not, in the first instance, producers. Standing apart from both the manufacturing process and calculations of profit and loss, they were able to engage in objective research. And being set free from all demands of a practical sort, they turned their attention to the study of nature.

Well grounded in the rhetorical learning and logical thinking of their tradition, these intellectuals of the early modern West did more than classify in the documentary style. Intent on systematizing and making general statements, they began to fashion a universal body of knowledge from the fragmentary information and guarded traditions of the workshop as well as from the voluminous literature of the high tradition from the Greeks to the Schoolmen.

Experiential knowledge of things acquired through observation is from its inception governed by the objects at hand. Experiential knowledge acquired in a cold climate is obviously going to be quite different from knowledge acquired in a tropical one. In other words, this knowledge lacks universal reference. In regions other than the one in which it was generated, it can only be used in an approximate fashion at points where the phenomena in question happen to be similar. The uniqueness of modern science lies, in contrast, not in its geographical origins but in its method. It is not prescribed by the objects it studies. Empirically minded scholarship dedicated to the acquisition of useful knowledge through observation and experience has appeared at different times and places throughout the world, but modern science is unique in having a universally applicable method.

Older modes of learning rested on theoretical foundations that contained a large admixture of extra-scientific notions peculiar to the region in which they were formulated. The Greek theory of the Four Elements, the Indian doctrine of Mount Sumeru, and the Five Elements of the Chinese are all closely linked to and identified with the life and culture of their regions. Prior to the emergence of modern science one might say that the minds of scholars were ruled by geographic accident. Those born in China became Confucian and adopted the Five Elements theory, while Westerners were Christians who accepted the notion that there were only four elements.

When the 17th-century Japanese Confucianist Mukai Genshō (1609–1677) prefaced a popular interpretation of the Aristotelian cosmology that had entered the country during the previous century, he explained:

> Men who write horizontal script and eat with their hands believe that there are four elements. They inhabit the lands of the Western ocean and India. Those who write vertically and eat with chopsticks hold the opinion that there are five. Their countries are China and Japan (*Bummei Genryū Sōsho*, 1914).

A generation later Nishikawa Joken (1648–1724) could write:

> I do not know much about the Four Elements and the Four Properties of the West, but in this part of the world we have the Five-fold Cycle (*of the primary elements as they successively produce and destroy one another*) and the Six Atmospheric Influences. Since altogether we have eleven as against their eight, are we not more knowledgeable in these things than they? (Nishikawa, 1712).

The contemporary student of comparative cultures demands more subtle criteria, but in suggesting that the problem was wholly one of number, Joken reminds us of the idiosyncratic character both of the figures themselves and of the choices that lay behind their rise to dominance in their respective cultures. In ancient China we read of four and six in addition to five elements, and in the Anyang "oracle bones" there is even mention of nine. The idea that there were five elements seems to have won out, according to Shinjō Shinzō (1928: 640–43), only after the belief that there were five and only five planets became fixed during the 4th and 3rd centuries B.C.

But even though it was developed in the West, one speaks of modern science as "modern" and not as "Western" because its method has been transplanted into non-Western countries and employed on a large scale without revision. When we say that the particularity of modern science lies in its universality, the words seem to have a contradictory ring, but this is because we confuse the idea of universality with the notion of timeless truth. In saying that modern science is universal, we simply mean that there are no regional limitations on its applicability to the phenomenal world. Yet this same method (the origins of which include an element of chance) possesses such marked individuality that its beginnings defined the beginning of a new historical epoch. Before pursuing this aspect of the matter further, however, let us first set forth the main components of this unique individuality.

Amidst the upheavals of the 17th century, all sorts of potential paradigms made their appearance. By the 18th century, however, the work of Galileo, Kepler, Descartes, Boyle, and others had been amalgamated in the Newtonian paradigm. All these men shared a mechanistic way of thinking. Laplace in particular sought to explain the whole physical world in terms of particles and the forces that operated between them. These were notions quite independent of the place where they originated. They possessed, in other words, a universality that transcended regional differences. Moreover, as concepts, the utility of "particles" and "forces" was not limited to a single field of research. Astronomers, physicists, chemists, and, somewhat later, biologists and even students of society came to feel that everything could ultimately be explained in these terms. One by one, this mechanistic view of nature pervaded the academic disciplines, until it seemed that it would invade all.

A theory of matter in which all properties were reduced to those of particles had been advanced by the Democritians in ancient Greece, and a mechanistic view of nature had appeared in the Hellenistic period with Strato and Archimedes. In the Middle Ages, however, both were buried beneath the dominant Aristotelian tradition. It was not until the Renaissance that they were rediscovered and set in place as a nucleus around which the edifice of modern science was to be built. But the presence of the Aristotelian paradigm is part of the story. Modern science emerged in revolt against a venerable opponent which provided both a focus for criticism and grist for controversy and discussion. Had this been a revolution against the Confucian paradigm, it would have taken quite a different form. With its general lack of interest in the sciences, Confucianism in fact proved quite unable to provide a material basis for the rise of modern science.

The failure of Confucianism in this respect is often contrasted with the prominence of logic in the Aristotelian tradition. Yet it is hard to find a historian who believes that this aspect of Aristotelianism played a positive role in the 17th-century Scientific Revolution. The period between the 15th and 19th centuries was actually a dark age for logic (Kneale and Kneale, 1962). But the medieval Scholastics' interest in Aristotle went beyond logic. If his logic was their initial preoccupation, they were also concerned with his *Physica*, and it was in this latter incarnation that he served as a sparring partner for modern science. It is in the study of nature that historians note a shift from traditional logic to experimental approaches inspired in part by the practices of artisan-technicians, and to Platonic mathematics.

Experiment and mathematics do lead into unknown worlds. It is chiefly experiments and attempts to express experimental results quantitatively that distinguish the modern scientist from the student of things who accumulates knowledge for practical ends.

In the pre-modern world, astronomy was the most highly respected of all the sciences. From East to West it was the exact science *par excellence*. Yet even though one thinks of it as being at the forefront of the 17th-century Scientific Revolution, its subsequent performance was lackluster. Eventually it fell by the wayside and became a small and comparatively insignificant field in the vast scientific enterprise. It failed to keep pace with burgeoning growth in many other fields because, unlike modern physics, chemistry, and (more recently) biology and geology, which adopted experimental methods that permitted them to both alter and simulate natural conditions, astronomy remained dependent on passive observation.[2] When the Jesuits came to the Orient in the 16th and 17th centuries, the astronomy they brought with them was not incomparably superior to what they found in China and Japan, despite the fact that the West was in possession of the heliocentric scheme and had, in the telescope, a revolutionary means of observation. Mechanics, modern chemistry, electromagnetism, and the other modern sciences introduced by the Europeans, however, were strange new species of learning that had no counterparts among the practical studies of the Orient.

In an experiment one does not know but what he is going to get a devil or a snake. Of course experiments are set up and predictions made within the framework of a given paradigm. Yet once an experiment has begun, there is something about it that keeps one spellbound until the results are in. It was for this thrill that men once squandered vast amounts of effort on what is now considered the pseudo-science of alchemy (a theme which is, incidentally, vividly treated in Balzac's novel *La Recherche de l'absolu*). Yet experiments conducted under the aegis of mechanistic ideas were something new. If any attempt to manipulate and observe objects in nature is thought of as an experiment, then such things were not wholly unknown before modern science. Occult objects generated particular concern in the black arts of Europe, while the Chinese displayed avid interest in magnets and were fascinated by thunder and lightning. But in the magical traditions, experiments were exclusively concerned with the abnormal and the extraordinary. When a given experiment was shown to be mechanically repeatable, scholars lost interest in pursuing the matter. In contrast, experiments guided by mechanistic theory were designed with predictability and control in mind so that predic-

[2] Some consider that astronomy entered the experimental stage with the appearance of artificial satellites, but this development should be seen first of all as the introduction of a new revolutionary method of observation.

tions might be confirmed, and quantitative precision increased.

But what of mathematics, the other weapon of the new science? Mathematics, being characteristically concerned with measurement, was of that practical type that Francis Bacon called the handmaiden of science. Japanese *wasan* was to become a dilettantish amusement. Neither had sufficient philosophical bent or ideological pretension to encourage the claim that "nature was written in mathematical figures." Of course it is not obvious that nature is expressible in mathematical terms; in fact, there would seem to be any number of things that cannot be expressed mathematically. In this sense, the mathematical view of nature was simply a kind of faith—some might even label it a superstition. Nevertheless, this conviction has given modern science a highly efficacious method, one of its distinguishing characteristics.

Mathematics, in contrast to what is taught at school under that name, can be a means by which the imagination leaps out of the confines of logic and factual evidence. To Neo-Platonists like Kepler and Galileo it offered an invitation to get away from the encumbrances of everyday life and take flight into the abstract, ideal world of numbers. When they sought to grasp planetary motion in mathematical terms and debated anew the problem of the dimensions of space, the laws at which they arrived could never have been obtained through the accumulation of factual evidence alone. They emerged from a vision of ideal conditions that appear only in a mathematically ordered world.

Happily wed under the aegis of the mechanistic paradigm, experiment and mathematics—sensation and ratiocination—were, in tandem, the two major concepts that set thinkers off down the track of normal science. They continued to propel the locomotive of modern science long after they had lost the ideological significance they once held for the paradigm builders and had become mere techniques. In order to understand just how effective these concepts were and what they contributed to the paradigm of modern science, let us take a comparative look at some of the older paradigms.

The first of the ancient Greeks associated with the notion that there were four basic elements was Empedocles (ca. 493–433 B.C.). He himself did not formulate the sister concept of the Four Properties (Dryness, Wetness, Cold, and Heat), but this idea influenced the Pythagoreans as well as certain medical groups, and by the time of Aristotle the notion that there were Four Elements and Four Properties had become settled opinion (Burnet, 1920). The number of elements and properties was not, of course, the only dimension of Greek theory. The Four Elements (earth, water, air, and fire) were viewed as being irreducible and substantial, the basic stuff of which the material world was formed. The Chinese, on the other hand, conceived of their five *hsing* (wood, fire, earth, metal and water) more functionally. Though commonly translated "elements" *hsing* might be better rendered as forces, energies, or phases that operate in a cosmic cycle. But whether it was Four Elements or Five Phasal Forces, the choices of the ancients involved, at the phenomenal level, a fairly high degree of arbitrariness—more at any rate than the particles and motion of modern science. Yet once these notions became "fundamental truths"—once they came to serve as paradigms—all phenomena were fit into the paradigmatic framework, or else the paradigm was stretched to fit the phenomenon. And the more "fundamental" and "abstract" they became, the more they took leave of the phenomenal world and became lofty doctrines that could no longer be checked by observable facts. This tendency was particularly strong in China. Although Yin-yang Five Elements theory did become an underlying principle of astronomy, medicine, materia medica, and all the other sciences, there were many cases in which it lost contact with the phenomenal world and ended up as mere decorative ornamentation. Combination with and subordination to Confucian ideals of political and moral leadership in the Han period also helped seal the fate of what was originally a statement about the operations of nature. How far the doctrine moved away from the natural world is illustrated by the Ch'ing reaction to Western reports of the discovery of Uranus. The apparent fact that there were now six planets was accepted, but the doctrine of the Five Elements had become so completely divorced from its connection with the Five Planets of the heavens that its status as a basic principle of natural philosophy does not seem to have been affected at all.

A similar tendency can be seen today in the type of formalism that brandishes paradigmatic formulas but rejects the phenomenological feedback and corrective checks provided by the phenomenal world because the formulas have been elevated to abstract dogmas. Paradigms simply cannot function properly when theory is so rigid as to be effectively immune to phenomenological constraints. But excessive flexibility also limits paradigmatic function. The structure of the Yin-yang Five Elements scheme was so loose that it managed to incorporate even the most empirically questionable phenomenon in one way or another. As a result there were no crises born of contradictions between the paradigm and phenomena that it was unable to explain; nor did anyone conceive the novel idea of pitting the paradigm against the phenomenal world in order to verify it. Indeed the Yin-yang Five Elements theory consistently performed only the most elementary function of a natural philosophic doctrine: satisfying the human need for a sense of cosmic order.

In making use of mathematics and experimentation, the new paradigms of modern science sought to be responsible to phenomena. If theory and experiment or observation did not agree, one of them was held responsible and revised. This process of making theory responsible to phenomena led in the case of celestial mechanics from the problem of two bodies to the problem of three and then to the *n*-body problem, stimulating the training of large numbers of modern scientists and sending them off on a headlong dash down the Newtonian track.

But there were certain modes of learning that waned in the wake of this explosive new orientation and came to be known as pseudo-sciences. In the common view of historians, these pseudo-sciences were mixtures of fact and superstition that existed before the sciences asserted themselves: before chemistry there was alchemy; before astronomy, astrology; and so on. Yet the fact is that both alchemy and astrology (especially horoscope astrology) actually had elaborate paradigms. Nor do such organized bodies of knowledge simply appear where there is no prior empirical accumulation. Still, neither alchemy nor astrology can be adequately understood simply as scientific activities that lacked modern paradigms. Chinese alchemy was much more interested in the techniques of achieving spiritual perfection and immortality than in the mundane goal of changing base metals into precious ones, while astrology aimed at enforcing the emperor's moral fitness to rule—goals which differ from those of modern science. At the same time, alchemy and astrology were attempts to rationalize empirical techniques and accumulate scientific knowledge by applying a certain conceptual framework (belief in an elixir of immortality or in correlations between celestial omens and human fate). For instance, when Ptolemy's *Tetrabiblos*, which embodied the paradigm of the horoscope astrology of the West, discussed the influence of celestial phenomena on the earth, it did so on the basis of the achievements of the mathematical astronomy of its time, including the notion that the movements of the planets had a lawful character. Thus it would be more correct to speak of alchemy and astrology not as pseudo-sciences but as the sciences of longevity and fate prognostication—modes of learning of a kind that are simply different from (modern) science, less powerful but less restricted, since they did not exclude moral experience.

Why then have they come to be known as pseudo-sciences? In the final analysis it is because they could not be subsumed beneath the paradigmatic categories of modern science. Modes of learning like the Aristotelian paradigm that are customarily accorded a legitimate place in the history of science tend to be those that encountered contradictions and anomalies with which they were unable to cope and as a result gave way to new paradigms for which they served as a point of departure. The pseudo-sciences, however, lost their intellectual vitality and died out without being effectively challenged by a new paradigm that might have replaced the old and transformed the tradition. Neither alchemy nor astrology could be substantiated by modern science, but they could not be disproved either. What happened was that people began to perceive later generations of practitioners as unsuccessful and ceased to be interested in their investigations. The two traditions simply died on the vine.

Scholarly paradigms do not all seem to be subject to revolutionary change. They do not all generate anomalies that lead eventually to scientific revolutions and the establishment of new paradigms. This model fits the physical sciences, but it fits poorly or not at all in the human and social sciences. Alchemy and astrology were not the only modes of learning in which *rigor mortis* set in when they ceased to generate attractive problems. In the Middle Ages, Scholasticism was everything. Confucianism enjoyed a similar and even longer reign. In their days of glory, it must have seemed as if they would continue to grow and develop forever. Yet they lost their appeal and became frozen monuments of learning that today are of interest only to historians.

The Mechanical View of Nature

There is no guarantee that modern science will not meet the same fate as Confucian and Scholastic learning, no assurance that its progress will continue forever. To be sure, modern science has a spatial universality that makes it neither "Eastern" nor "Western." This is why it is transplantable to any part of the globe. Nevertheless, temporally speaking, its methods remain products of a particular scholarly movement that took the world by storm at a particular point in history.

This movement had at its center a science of mechanics that was ready to explain everything in terms of matter and motion. In the mechanical view of nature, things like color were secondary attributes to be reduced to the wavelengths of spectral analysis. However, sensation (including the sense of color) is a psychological phenomenon, not reducible to the categories of mechanics or physics. Yet in selecting mechanics as its model discipline, modern science built a peculiar hierarchy of discrimination in the academic domain.

Historically speaking, modern science began by tackling problems that lent themselves to resolution by mechanical methods. The solutions were quick in coming, thick and fast, like water flowing into and filling up low-lying areas when a dam bursts. But small hills always remained. Earthquake prediction and weather forecasting—two areas in which the Chinese left voluminous records of phenomena—proved recalcitrant. The study of culinary tastes and smells was long neglected. More generally, a hierarchy of disciplines emerged in which the status of any particular discipline or field of inquiry was roughly determined by the degree to which the problems it posed were amenable to mechanical solution.

The Exclusivist Character of Modern Science

With its all-surpassing tools of number and measure, modern

science stood in judgment on the rest of the academic world, distinguishing between science and non-science and repudiating the latter. In Chapter 2 I described the difference between traditional Chinese and Western attitudes toward the anomalous. The Chinese included all kinds of irrational things in their purview, classifying and finding a place for them. The Western tradition held fast to the lawfulness that was believed to underlie the phenomenal and tended to close its eyes to things that resisted expression in legalistic terms. Modern science took up this aspect of the Western tradition with a new aggressiveness, denigrating what did not follow its methods.

There are mountains of Chinese materials describing popular medical techniques which, from our present vantage point, seem of highly dubious value. When Chinese scholars came to put together the canons of classical medicine, they included a careful selection of those techniques that were currently in use. Thus popular medicine proved to be a source of materials for classical medicine without the two being seen as antagonistic, for the pliable structure of the Yin-yang Five Elements paradigm of Chinese medicine was not given to the drawing of hard and fast lines. An abundance of popular medical techniques also existed in the West, but modern medicine, unlike its Chinese and Western predecessors, judged them unscientific and refused to accept a state of coexistence. This is the exclusiveness of modern science.

If mechanics occupied the central place among modern scientific methods, physics, chemistry, and biology were situated around it, in that order of status. Problems of society stood at the periphery (see Figure 4–1). To speak metaphorically, each of the incipient paradigms that emerged during the 17th century acted as a kind of vortex. A list of paradigmatic works which emerged from within the mechanistic view of nature alone would include those of Kepler, Galileo, Descartes, and Boyle. In describing the formation of modern science, science historians have often succumbed to the temptation to bring all these vortices together in a single line of evolution. In truth, however, each incipient paradigm was distinctive enough to have given rise to its own line of research had it continued to develop independently.

In the interim, however, the Newtonian vortex appeared. Churning with even more possibilities, it swallowed up the Gali-

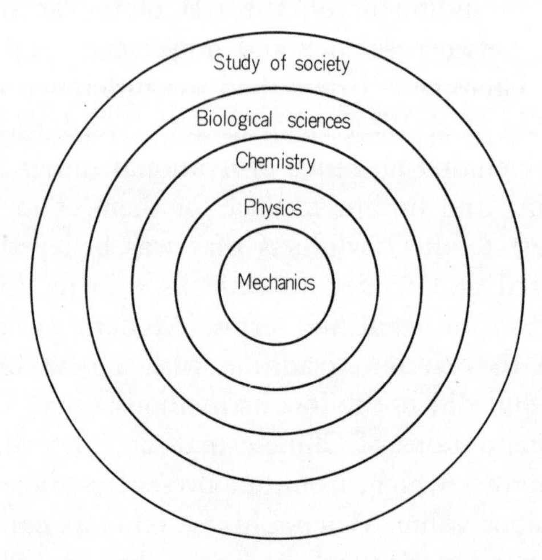

Figure 4–1

lean and Keplerian vortices, gathering groups of supporters as it went. Initially, the rival Cartesian vortex loomed larger, but in the 18th century the Newtonian paradigm was to emerge triumphant. Newton's vortex had originated in the fields of astronomy, mechanics, and physics, but as it grew it enveloped chemistry and biology, bringing into being a modern science that would for a time seek a mechanical explanation for everything. Of course chemistry and biology had their own small vortices around Lavoisier, Mendel, Darwin; and these fields continued to produce and develop fairly elaborate paradigms that could not be easily reduced to physics, even after Newton. And modern modes of ordering experience that originated far from the mechanical center of modern science, like the Marxian, will probably always resist reduction.

Even today, in looking at the development of fields like molecular biology one may be cajoled into thinking that reductionism is entitled to make exclusive and absolute claims to truth, but this is simply because, in our historical experience, it has done better than other approaches. We should, however, by no means invest this fact with ultimate significance. The genesis and development of new vortices at a variety of levels in chemistry, biology, and psychology is of considerable import, and it just might be that a major new paradigm will emerge to replace the 17th-century mechanistic one. Nor is a major revolution within modern science that would radically alter the disciplinary divisions beyond the realm of possibility.

The final distinctive characteristic of modern science that I would mention here is a negative one: the absence of the Aristotelian notion of a final cause, the lack of teleology. The artisan's techniques on which Aristotle modeled his theory—the building of a bridge, the making of a desk, etc.—were from their inception linked to a specific purpose (Ravetz, 1971). But in modern science and its mechanical and mechanistic view of nature, purpose has no place. In the nature of things, bodies and particles are regarded as moving aimlessly about. Experiments are designed to control these movements for the moment, but in a broader sense, one does not know where "value-free," "non-teleological" research is going to go. More generally, the researcher has no idea where the normal science he practices might be heading. In short, inasmuch as it has a method but no end, there is always the possibility that modern science will run away with itself, using its methods recklessly to create an avalanche of violence.

From Books to Articles: Changes in the Form of Scholarly Publication

Inasmuch as China had printing as early as the T'ang and Sung dynasties, one might say that Gutenberg finally brought the West abreast of Oriental standards in written communication. But the difference (in both cost and speed) between block printing, with individual pieces of wood carved full of ideographs, and printing in an alphabet of letters that could be arranged and rearranged in movable type meant that in catching up, the West was also on its way to forging ahead. These differences were not so significant in the printing of classical texts, but the advent of newspapers and magazines made the advantages of movable type quite apparent.

Still, printing technology does not tell the whole story. The dissemination of technology is decisively shaped by social institutions, and one notes among the institutional developments that supported history's greatest scientific revolution two things that were simply not to emerge from within the academic tradition of the East: scholarly journals and learned societies. (The Han-lin yuan may be thought of as a learned academy, but it was not responsible for disseminating learning, even among its members. It was an Imperial advisory and drafting bureau quite remote in spirit from the voluntary societies so important in Europe.)

Europeans created academic societies to exchange ideas, and printed journals in order to make the results of their discussions more widely available. These things would not have been impossible in Chinese society. That they first appeared in the West has less to do with the Asiatic mode of production or the peculiar features of a self-sufficient agrarian society ruled by a civil bureaucracy than with the possibilities offered by Western societies in the heyday of mercantilism. Traders went off to the far corners of the globe. When they returned to Europe they wanted to share their new knowledge, and they were met by large audiences eager for exotic news of the larger world. Intellectual salons formed among the more curious, and it was by no means unusual for these groups to develop into learned societies. The old intellectual elite, the Schoolmen, were left behind by educated laymen with the opportunity to pursue their own interests.

If curiosity pulled, the need to "keep up" in a constantly changing mobile society also pushed those in European intellectual circles to value new developments in the world of knowledge. The writings of Francis Bacon captured this atmosphere and gave it concrete expression in an intellectual program that was to provide the guiding principles for a diversity of 17th- and 18th-century societies concerned with the exchange of information and ideas.

Prior to the invention of the journal, scholarship was largely confined to the tasks of annotation and compilation. This was, in fact, true of Chinese scholarship until quite recently. But with the geographical discoveries and the experimental science of the Renaissance, there was an increase in new knowledge that this form could not accommodate. At first it found expression in notes or supplements to texts of an earlier age like Ptolemy's *Geography*, one of the several works of Aristotle, or the astronomy textbook of the medieval university, Joannes de Sacrobosco's *De Sphaera*. But so much data was accumulated that *De Sphaera*, for instance, which began as a thin, quite mediocre little book, became with Clavius' commentary a voluminous, specialized work that bore little resemblance to the original. The commentary was becoming a less and less serviceable format.

Writing a book to announce every new discovery was a burdensome and impractical task. Books were ordinarily of substantial length, and the classic-plus-commentary form made it difficult for the reader to determine just what the new knowledge was and where it was located. When the exegesis included the opinions of everyone who had ever said anything about the subject, the commentary grew to be much more extensive than the text, and the reader grew correspondingly more perplexed.

The response to this dilemma was a mode of expression that dispensed with the original text (taken to be common knowledge) and succinctly recounted only what was new. Before the dissemination of printing, new knowledge and information was sold in the form of handwritten dispatches, but these were expensive and available only to the aristocratic class. The journal put this knowledge and information in print.

Behind this shift from books to journal articles there was also an important change in scholarly style. Annotated commentaries on the Aristotelian corpus contained rhetorical learning that had taken shape as the founder and his descendants did battle with their intellectual adversaries. Seventeenth-century scientific research also tended to take this form, often being more intent on destroying the arguments of opponents past and present than in looking at nature as an objective phenomenon. Since it takes time to compose something as large as a commentary, 17th-century scholarly opinions were frequently exchanged in letters. But these exchanges were also extremely polemical, for handwritten accounts were usually written with a specific opponent in mind. The modern scientific article as we know it today has been largely stripped of all that polemic and comes attired in the garb of "objective scholarship." A polemical style that attacks particular individuals cannot be successfully employed when printing expands potential readership to a large and anonymous group. One must write in a way that will be acceptable to large numbers of persons—although, as I will note later, the objective dress of the modern scientific article can be deceptive.

Early Scientific Journals

The bulletins and periodicals of modern print journalism could not exist without an established postal system. The merchandising of books and printed materials and the relation between distribution and censorship systems and the communication of learning are other factors that must be understood if we are to fully comprehend the circumstances of intellectual life during this period. Unfortunately, however, the history of science and learning has scarcely begun to explore these areas. Thus we must be content to note that journals came to be issued in 17th-century Europe under new social conditions that were generally favorable to the communication and transmission of knowledge.

Before the modern scientific journal made its appearance, there were two sorts of publications that performed roughly the function for which the journal was destined. The first of these was the book catalogues and advertising sheets put out by publishing houses or book markets (Frankfurt and Leipzig are famous examples), most of which contained synopses of the books they listed. (It may be observed in passing that many more titles were published in Europe in the years between the introduction of printing techniques and the 18th century than is the case today. In an age when there were few other places for advertising and book reviews, this may have had something to do with the catalogue publisher's need to acquaint the reader with the content of the books he listed.) A second type of publication was the recorded proceedings of learned societies, which to some extent replaced individual exchanges between scholars. The early journals served as both book catalogues and proceedings, but the former function was more popular and more prominent. In the 17th century in particular, scientific journals were in effect reviews of and introductions to books (Kronick, 1962).

The first scientific articles were not the showcases of scholarly accomplishment one finds today. They were miscellaneous scientific reports designed to provide preliminary information about new developments and discoveries that would otherwise have remained unknown until the time-consuming task of producing a book had been completed.

The scientific journal itself is usually said to have had its inception in two publications: the *Journal des Sçavans*, first published in Paris in January 1665 (publication was halted at the time of the revolution, and the journal went out of existence in 1792), and the *Philosophical Transactions*, which appeared two months later in London and still exists today. If the former may be seen as the original general intellectual magazine, the latter can be called the prototype of the specialized scientific publication. The stated aims of the *Journal des Sçavans* were as follows (Brown, 1934):

1. To provide a list of books and a brief introduction to their content.
2. To report the deaths of well-known figures and summarize their achievements.
3. To report on experiments in physics and chemistry which aim to explain natural phenomena, on new discoveries in science and technology, on useful instruments and experimental apparatus, on the ingenious inventions of mathematicians, on astronomical observations, on climatic phenomena, and on new discoveries made in animal dissections.
4. To report on major events at the court and the university.
5. To report on the present state of the intellectual community.

Summaries and synopses occupied a major portion of these early journals, and were, indeed, among their distinctive features. As an intellectual operation, summarizing requires one to generalize on the basis of experience, observations, or available data, and it may be comparatively observed that the Orient—where such publications did not exist—was as lacking in the generalizing impulse as it was in the logical dimension. There would seem to be a profound relationship between generalization and all kinds of logic, but my point here has to do not so much with the deductive logic of the Schoolmen as the inductive, Baconian operation in which one arrives at general axioms and theorems by taking up the diverse phenomena of nature and subjecting them one by one to a critical examination. The characteristic Chinese mode of academic expression—editing and annotating classical texts—had no place for this kind of generalizing, but it arose quite naturally in the Western journal format.

The *Journal des Sçavans* took a critical stance toward church and state, and it quickly fell prey to periodic acts of suppression by the authorities, but the *Philosophical Transactions* was much more specialized in its orientation and orthodox in its politics. It was a collection of letters to the editor and articles contributed in response to questions raised therein. To the extent that this format perpetuated an older tradition in which new scholarly information was put into individual correspondence, one can also see here the stylistic predecessor of our modern-day scientific treatises.

These early journals did not emphasize authorship. The articles were not signed, but the work of individual researchers was publicly recognized in hearsay form ("we have heard that so-and-so is now conducting such-and-such an experiment"). During the period in which research was communicated in individual correspondence, there were constant quarrels about who had discovered what first; but with the advent of journals and their umpiring editors, the acrimonious disputes of the kind that were carried on between Newton and Hooke and Newton and Leibnitz became rarer.

Types of Research Publications

The transition between the latter half of the 17th century, when alternative forms of research publications appeared, and the time when the academic journal as we know it today consolidated its position is analogous to the emergence and maturation of a new scientific paradigm. Several prototypes appeared, and one drove out the others and underwent further elaboration and refinement. In the 17th and 18th centuries one can note at least three such prototypes: 1) the independent journal, 2) the transactions of learned societies, and 3) the published university thesis.

What then was the role of the journal, the learned society, and the university in the creation of a new academic tradition? The university was by far the oldest of the three. Learned societies and journals flourished side by side in the 17th century, though it can be argued that the former appeared first (the Accademia Pontaniana was founded in 1433). The Accademia dei Lincei was founded in 1603, and the *Journal des Sçavans* was first published in 1665. The significance of their contributions to the scientific revolution, however, seemed to vary inversely with their age.[3]

When someone acquires new knowledge, his first thought is to tell it to others. If he stores it in his head, it soon loses its freshness and he eventually loses his desire to communicate it. Thus the sooner he is able to pass it on, the better. Promptly reported, it stands a better chance of gaining the attention of others interested in the subject and—should it contain the seeds of a new paradigm—leading to the formation of an advocate group which will eventually develop into an academic society of professional specialists.

An academic society is a group of persons who support a mature paradigm. If the members of the society are to train successors who will continue the tradition, they must wedge their way into the university—that is, into institutionalized education. The journal consistently spearheads the paradigm's advance, and when academic societies come into being, the paradigm acquires a place within them. In the end it attains full citizenship in educational institutions, where group replenishment can be systematically carried on. With this, the process by which a particular mode of learning becomes established in organized society is complete.

One can see a tendency toward this general pattern of growth and change in the scientific publications of the 17th and 18th centuries. First there were the journals—most of which aimed chiefly at disseminating knowledge and educating their readers. Articles were not necessarily selected for publication on the basis of scholarly merit. In most cases, these journals were run by a single individual who had responsibility for everything from editing to sales. He was always looking for whatever new knowledge and information he could find to publish, and he at least hoped to make a profit. "Knowledge is power," went the Baconian dictum, and these journals were faithful to it, inspired by the Enlightenment goal of liberating mankind from ignorance through the dissemination of knowledge. As they picked up readers among the bourgeoisie, they began for the first time to acquire a stable economic base.

[3] Academies can be found as far back as ancient Rome as well as in the medieval Islamic world, but these bodies were primarily secret societies or government advisory groups. Modern academies performed a different function.

The readers of these journals were not specialists. Specialists in the present sense of the word did not, in fact, exist until the 19th century. To be sure, there were the old groups of specialists—clerics, lawyers, physicians, and teachers—trained respectively in the theology, law, medical, and philosophical faculties of the medieval universities. But the old guilds had little to do with the new knowledge. It was rather individual clergymen, teachers, and physicians who, in more or less equal numbers, transcended the confines of their professions and shared their ideas in the learned societies and scholarly journals of the 18th century.

In the 18th century, journals had not yet come to monopolize the publication of original research. Some learned society "transactions" were also scientific publications. Yet there was normally a considerable lapse of time between the presentation of a paper at a society meeting and its appearance in print. The most famous and prestigious of the transactions, the *Mémoires* of the Académie des Sciences, took from two to six years to publish an oral presentation. When the *Mémoires* came to be edited by committee, the circulation of manuscripts among committee members consumed even more time. Meanwhile, publication of the records of academy meetings was often put off until enough manuscript pages had accumulated to publish them in book form. As a result, it was not uncommon for even a first-rate scientist to send his manuscripts first to the more popular, privately edited journals when he wanted to stake his claim to a new discovery, invention, or theory. Under these circumstances, the transactions of learned societies assumed an authoritative presence, while the independent journal maintained the type of free, *avant-garde* atmosphere that encouraged scholars to write without undue anxiety about their fallibility.

The economic base of these privately edited popular journals remained weak, however; and when an editor died or lost his patron, his journal might simply vanish. Journals were also victims of political pressure. Even publications in medicine and the natural sciences sometimes incurred the wrath of the authorities. Governments often tended to be suspicious in principle of these new journalistic endeavors. The proceedings and transactions of academic societies tended to be much more respectable, for they were in the custody of ongoing organizations.

Data compiled by Kronick (1962) on 1,052 periodicals that appeared through 1790 include the following figures concerning their relative stability:

Duration	*Academy Bulletins*	*Independent Journals*
Five years or more	63%	31%
Ten years or more	46%	18%
Thirty years or more	21%	4%

After 1790, however, independent journals and proceedings and transactions of academic societies suffered a common fate. Buffeted by the great wave of specialization, they both fell into steady decline. Specialized journals had begun to appear about the middle of the 18th century, led by medicine and then agriculture. Physics, biology, and technology also acquired specialized journals in the 1890s, and the 19th century ushered in an age of specialization.

When this happened, the general intellectual magazine ceased to function as a professional publication. This was also true of the proceedings and transactions of the older learned societies. With the 19th-century rebellion against the Royal Society and academies in other nations by scholars who sought to found specialized academic associations in their several fields, academy publications that did not specialize were left to gather dust on the shelves. With a few exceptions (like *Science* in the U.S.) these publications have meaning only in underdeveloped or small countries where research is not far enough advanced to warrant specialized journals. Here they stand as a symbol of national independence and continue to set standards for the intellectual life of the nation.

According to Kronick's statistics, Germany accounted for by far the largest number of journals published through 1790, representing 62% of the total as against 11% for France and 7% for England (see also Garrison, 1934). Germany may have been the home of Gutenberg, but this extraordinary set of figures can hardly be explained in terms of some innate German love of the printed word. It is best accounted for in terms of the decentralized German political order. In neighboring France, there were local academies that modeled themselves on Paris, but these bodies tended to operate like branches, sending their reports to Paris rather than publishing their own. The German states, however, lacked a central nucleus. Where local culture flourished there were journals which in turn supported the cultural life of the local city.

Because these journals were so numerous, though, they tended to command only limited financial resources. Many of them went out of business shortly after being founded. When one takes the shorter average life of these German publications into consideration, the number of German journals in existence in any given year expressed as a percentage of the whole drops below 40% of the total.

Nevertheless, Germany continued to produce more than its share of journals throughout the 19th century and does so even today. Many of these—throughout Europe, but particularly in Germany—are still edited by individuals. Though officially periodicals, most of these publications are in fact occasional, appearing whenever sufficient manuscripts accumulate and funds for publication are found. Naturally, the character of these journals is heavily colored by the personality of the professor–editor. When one editor retires, his successor may institute completely new editorial policies, sometimes even changing the name of the publication.

In the United States, things moved in the opposite direction. Scientific journals were modernized and research was brought together in one major publication in several special fields. Corporate editing and selection of manuscripts also distinguished American journals. Standardization resulted in higher overall quality, but individuality suffered.

The story of that venerable flagship journal of the history of science, *Isis*, is a case in point. Long sustained by the personality (and funds) of its founder, George Sarton, it became after his retirement the official organ of the History of Science Society in the United States, under whose auspices it acquired an editorial committee and a slick new format. Since this transition, articles have had to meet rigid standards, and those of obviously inferior quality no longer find their way into its pages. Yet somehow the journal seems to have lost the peculiar appeal it had in the days of Sarton, when diamonds were often found in the rough. The journal has done its part in upholding professional standards, producing issue after issue filled with model articles. The unusual, the offbeat, the extraordinary have disappeared, however.

Once this type of journal has been created, scholars write to fit its specifications. Scholarly journals tend to become exclusively devoted to the advancement of normal science, and to lose those qualities that gave them a role as the voices of advocate groups during the period of paradigm formation.

Did other types of publications act as a stimulus to and a medium for the scientific revolution? This question admits of no easy answer. The role of the university thesis, the third of the three main types of scholarly publications in the 17th and 18th centuries, was negligible. Before the 19th century, published theses were particularly numerous in the field of medicine and in the German states, but they tended to be formal exercises in which content was secondary. They also were not generally available. Little more than repositories of knowledge, they were quite far from the frontiers of scholarship and the front lines of scholarly development, and much the same can be said of them today. In countries where specialized academic associations have yet to be established, university bulletins provide a significant outlet for the publication of academic research. Here their publication affirms the university's existence just as academy bulletins underscore the existence of the state as a sovereign entity. Yet inasmuch as these publications are not for sale to the general public and are otherwise very difficult to get hold of, there is little chance that they will come to the attention of the specialist. They can hardly function as media of scholarly communication, for the scholarly enterprise clearly transcends the confines of particular universities.

To continue to publish specialized treatises in university bulletins in a country like Japan, where associations of specialized professionals are well established, is a perverse act that only serves to hamper researchers in their attempt to stay abreast of what is going on in their field. The amount of money each university now spends in putting out a journal to advertise itself is quite insignificant. But scholarly journals might be appreciably strengthened if these funds could be pooled and expended on one or two journals.

The Modern Scientific Article

The more a particular paradigm acquires a stable form, the

greater is the functional division of labor between the paradigmatic classic (that is, the textbook) on the one hand and journal articles and reports on the other. Today mathematics leads the way in this respect, followed by the descriptive sciences and then the social sciences and the humanities.

Before Newtonian mechanics was sanctioned by academic society, the author of new research based on Newton's work had to begin by stating his own interpretation of Newton. Unless he proceeded in this fashion, the reader would be unable either to understand his work or to appreciate its significance. But early in the 18th century, a revised, annotated edition of the *Principia* became available. Further interpretations and elaborations were added, in which the names of Newton's supporters were massed to lend support to the Newtonian paradigm and defend it at crucial points from attacks by detractors. Reading these publications gives us a vivid picture of the intellectual controversies of the period. By the latter half of the 18th century, however, Newtonian mechanics had become axiomatic. Historical, episodic impurities were removed, and it was restated in a more logical fashion in modern-style textbook form. One no longer senses the enthusiasm of an advocate group eager to advocate and defend a particular position. Instead, theory and application are straightforwardly and efficiently presented so that both can be acquired with a minimal expenditure of time and energy. The name of the founder himself is no longer regularly used, being replaced in references to his science by technical terms like "mechanics."

Once a mechanics textbook appeared, a member of the Newtonian paradigm advocate group no longer found it necessary to begin with Newton. A scientific article needed to contain only those things that were not in the textbook. In a modern academic treatise the researcher presents the reader with a precisely worded account of his own original research to be appraised and evaluated without explicit reference to the paradigm.

But once articles growing out of a single tradition of research are mass-produced, a single individual cannot begin to digest the vast amounts of information that result. He is in much the same position as the historian: forced to rely on articles and secondary materials because he cannot hope to read all the original sources.

The mature scientific journal can be seen then as a product of the modern merit system and the modern emphasis on efficiency: presentation of the results of one's work in the shortest possible time with the least expenditure of energy. Scientists have come to report on their work in an arid style from which all accounts of the research program and the trials and tribulations encountered in carrying it out have disappeared. This style may suffice as a means of communicating the results of scholarly research, but it has been stripped to a considerable degree of the vital capacity to isolate new problems and stimulate broad intellectual interest.

Whether, in the process of transmitting an academic tradition, a creative atmosphere has ever been successfully communicated through the printed word alone is a question that science historians have not yet seriously considered. My guess, however, is that it has not. The style of the modern scientific article permits one to convey only what one knows. This means that the reader gets no feel for what is not known. The scientific article is an efficient vehicle for the swift acquisition of existing knowledge, but it offers few clues about where to go next, little sense of direction for future research. The problems that arise when paradigms produced in one cultural area are transplanted into a remote and vastly different culture are of a similar nature (see Chapter 6).

The Future of the Scholarly Journal

The scientific journal, having served modern science so well since it first appeared in the 17th century, has reached the limits of its effectiveness. Developments in transportation and the quickened tempo of scientific research have made the annual or quarterly journal much less suited to conveying information about the latest research than it was even a generation ago. Weekly and biweekly journals have been created, at the cost of refereeing and the quality control it implies. More fundamentally, however, the decline of the journal can be traced to the industrialization of scientific research itself.[4]

When the methodology of normal science is set, the research guidelines specified, the research contracts signed, the funds forth-

[4] Jerome Ravetz discusses the distinctive features of "industrialized science" in his *Scientific Knowledge and its Social Problems* (1971). See also Chapter 5 of the present work.

coming, and the data produced like goods manufactured in a factory, there is no longer any need to publish one's research and place it before the intellectual community. It is sufficient if the results serve those who want the data or desire the know-how. Moreover, the results need not necessarily be preserved forever. If a more precise coefficient of expansion for copper wire is found, the previous one can be thrown away. Under these circumstances, oral communication and letters—media that antedate the journal, together with computerized data, have become the "live" sources of information (Price, 1967). Scientists on the frontiers of search today look almost exclusively to preprints and dispatches called "letters" for new information.

Some medium other than the journal is necessary if the swollen storehouse of information is to be kept under control. For work conducted within an ongoing tradition, this presents few problems, but paradigm-building cannot be handled by computers and data banks. In all likelihood, the journal will continue to have a role, for there are requirements of this process—particularly that of opening research to public scrutiny—met only by the older forms.

The Role of Learned Societies

In her classic study *Scientific Societies in the Seventeenth Century* (1938), Martha Ornstein argues that the need for learned societies arose as a result of the failure of universities to adapt themselves to the revolution in the sciences. Galileo and Newton were, of course, both university men, and there was the rare institution that was kindly disposed toward the rise of modern science: the University of Padua is often cited as a case in point. But when one looks at universities as a whole and as institutions, Ornstein's conclusion that they worked to check the 17th-century Scientific Revolution seems unassailable. When Galileo was carrying on the fight for Copernican theory, the icy stares of university-based scholars only served as another fetter round his feet. These men could not stand idly by and watch the undermining of the Aristotelian paradigm that was the basis of their authority and the source of their livelihood as academicians. There are several indications that Galileo's anger upon being sentenced was directed less at the Papal Office itself than at the groups of scholars who plotted against him behind the scenes. On the other side, his sympathizers seem to have included numerous painters, sculptors, and other *avant-garde* freelancers as well as members of the newly rising bourgeoisie.

Disagreement over when learned societies began is partly a matter of definition. Some historians go back to ancient Rome (Edelstein, 1963), but the standard history of science usually begins with the Accademia dei Lincei (1603–1630) of Rome and the Florentine Accademia del Cimento (1657–1667). Galileo was associated with the first of these societies; the second was home to members of his advocate group who worked after his death to complete and develop his experimental method. These societies were clubs or joint research groups in which membership was restricted, much closer to Pythagorean secret societies than to the academic associations of today.

If the modern academic association has a spiritual father, it is perhaps Francis Bacon. In his vision of Solomon's House presented in *The New Atlantis* (written in 1617 but published posthumously) he sketched a model organization that would serve as a guide for the founders of the Royal Society of London, the Académie Royale des Sciences, and other learned societies. In Solomon's House, Bacon divided the process of conducting experiments and "ascending" to axioms and theorems into ten operations, and envisaged joint research projects being carried out one after another in assembly-line fashion. Bacon's vision was then neither of a club nor of an open academic association, but of a national research institute much more suggestive of the Moscow Academy of Sciences.

Bacon died in 1626, but no one in the academic world would dance to his piping for at least twenty more years. Looking for the source of this inertia, one runs straight into the putative central nervous system of the academic world—the university. Although Bacon's opponents did not wage a journalistic campaign, the universities were dominated by those bent on maintaining the Aristotelian ivory tower intact.[5] But during the revolutionary confusion of the Cromwellian period, a group of scholars at Oxford signaled their opposition to the university leadership and emerged after the Restoration as the Royal Society. The simmer-

[5] Costello (1958) is one of the few works that deals with the old paradigm in its last days.

ing conflict between this group of adherents to Bacon's suggestive proposal and the university never broke into full-scale combat, but it continued well into the 18th century (Purver, 1967).

In a parallel and rival development, the Académie Royale des Sciences was founded in France. The two bodies presented sharp contrasts at many points, providing two distinct models for the many learned societies that were to follow.

The most important contrast between the two lay in the realm of their relationship to and attitude toward government. The Royal Society was definitively shaped by the traditional *laissez-faire* policy of the English, according to which government was not expected either to have a say in or to contribute money to science and learning. In France, learned societies accepted government supervision and control under the patronage of enlightened despots from their inception—Richelieu's creation of the Académie Française under Louis XIII, and Colbert's of the Académie des Sciences under his successor.

The Royal Society was then simply an association of men with common interests, financed through membership fees and journal sales. Its members were not government employees, and appeals for governmental assistance were made only for special projects such as an expedition to observe a solar eclipse. The "Royal" in the Society's name did not mean that it had been founded by the court but that it operated with a charter from the King. Election to membership was less a privilege than an honor. But the Society gradually underwent a subtle change in character, a change that was perhaps not unrelated to its original constitution. Although still intellectually vital in the days of Newton, in the 18th century the need for financial reconstruction and other factors led to the admission of many members of the aristocracy, until the Royal Society was more a staid establishment club than a place for significant intellectual activity. The emergence of new, specialized fields in the 19th century stirred the Society from its lethargy, and a belated attempt was made to expel non-scientist members and rebuild; but the age of specialization was at hand, and the classical learned society concerned with all the sciences was no longer a suitable place for the development of normal science.

In France, membership was limited from the beginning to first-class scientists. Government stipends enabled these scholars to devote themselves to research; but they also involved themselves in the administration of scientific and university affairs, dominating the entire academic world. The members of the Académie Française were called *académiciens*, while those of the Académie des Sciences were referred to as *académistes* and received larger stipends. Both bodies were literally the King's. Thus the French came closer than the English to Bacon's conception of a learned society as outlined in *The New Atlantis*.

In the beginning, Bacon's suggestions about joint research were taken seriously by both the Royal Society and the Académie des Sciences, and many such projects were instigated. Joint research conducted by the fellows of the Accademia del Cimento achieved significant results, but the French-English projects do not seem to have fared very well. As a rule, it was individuals who made the big, paradigmatic discoveries. Men like Huyghens found little meaning in such projects, and when the Académie des Sciences was reorganized in 1699 joint research came to a halt.

Learned Societies and the Social Order

Do learned societies belong on the side of change or the side of order? Professional groups formed around a common discipline or technical skill, such as modern medical associations, are clearly sources of power, and sometimes, when their interests are involved, they behave as if they were political societies. For this reason, they have, during the course of their history, at times felt the oppressive hand of temporal authority.

The notion that new scientific paradigms appear only against the background of progressive political regimes cannot be documented. Even so, a revolutionary new paradigm almost inevitably finds itself uncomfortable in the old regime, and eventually gravitates toward reform. The Marxian paradigm is diametrically opposed to most existing politicoeconomic orders, but, on a smaller scale, innovative paradigms in the natural sciences also collide with academic societies and established institutions of higher learning. Moreover, the basic character of modern science engenders in its practitioners a peculiar cast of mind, a set of professional ethics and social attitudes that differ from prevailing values.

First, modern science stresses universality. If Newton's mechanics is true for Anglo-Saxons, it should be true for Hottentots and Eskimos as well. Hence the modern scientist is of a cosmopolitan temperament and holds to standards of value that are blind to regional differences. A Japanese novelist may have a million readers in Japan, but unless his work is translated, he remains unknown abroad. In contrast, many scientists have a limited but approximately equal number of readers among specialists throughout the world—across barriers of language, nation, and race, which the universalistic logic and ethic of modern science willfully disregards. Moreover, in contrast to dialogical and disputative forms of expression, the printed scientific treatise is evaluated without reference to the author's age, sex, skin color, faith, faction, origins, or social status. This system of values is something new in the history of human society, and it has produced among scientists an egalitarian and individualistic ethic which is at variance with the hierarchical values of most societies.

Attempts have been made by Dorothy Stimson (1935), Robert Merton (1949), and others to show that the majority of the members of the Royal Society were Calvinists, sympathizers with the cause of Parliament, and generally more radical than the defenders of Anglicanism and Monarchy. They observed, furthermore, that these men were the bearers of a Protestant ethic which provided the underlying spiritual foundation for modern science. Lewis Feuer (1963) has taken issue with the statistics used by these authors and argued that a careful examination of the data even reveals the opposite tendency. It has also been observed that the men who were later to found the Royal Society sought to avoid political strife during the revolutionary Cromwellian period, meeting to talk about scientific matters in the hope of finding some consolation for their distress at political conditions. Seemingly aloof from the rough and tumble of politics, scholars came to prize modern science's commitment to objective truth and its value-free orientation. The scientists of the Royal Society were also in the "middle" in terms of their social stratum, being chiefly drawn from sectors of the populace between the feuding upper and lower classes. One can even say that modern science was an ideology of the middle class. Indeed, according to the testimony of Thomas Sprat, the first man to write a history of the Royal

Society, the most numerous single group of charter members were physicians—who were, on the whole, neither partisans nor opponents of the regime but aliens and refugees from all regimes and would-be regimes (Sprat, 1667). It is the prevalence of such men that explains why, after the Restoration, the Royal Society asked for and received the official recognition without which it would have been viewed with suspicion and have been unable to establish its authority.

As I noted earlier, the Royal Society was a voluntary elective organization that received little financial assistance from the government. This was not because its members rejected governmental assistance out of fear that it would restrict their freedom to do research. On the contrary, members of the society were always venting their dissatisfaction at the government's apathy and indifference toward science, holding up the French government's patronage in pointed comparison. Yet there was no general criticism of the government and no organized political action.

Across the channel in France, the scholars appointed by Richelieu and Colbert (at least those in the Académie des Sciences) were less often ideological advocates of science than non-political specialists who excelled in their particular art. An atmosphere of support for independent research existed in the age of Colbert, but after his death the Académie, as an advisory body of the Court, was chiefly concerned with utilitarian topics such as gunnery mechanics and the hydraulics of waterfalls in the Palace gardens. Although they were the recipients rather than the creators of initiatives in scientific research, few among them suggested separation from government control. They even felt it to be a point of pride to be an official body.

Thus the learned academy, unlike the journal, was not *avant-garde*. It was itself an establishment. Election to membership quickly became a status symbol—the first step in institutional closure to the outside world. With it began a process that has been repeated many times since: the fashioning of a refuge for the preservation of learning where, untroubled by stimuli from the world outside, scholars created a merit system among the fraternity that soon became deeply entrenched.

As a self-supporting association financed through membership fees, the Royal Society was to be the prototype of learned societies in the countries of the free world. The Society lacked money, but it cultivated the idea that election to membership was an honor. Insofar as new members were chosen by the existing membership, a quick hardening of the arteries was inevitable. While an interest in science continued to be a condition of membership, after large numbers of aristocratic gentlemen with only a casual interest in science were admitted to shore up finances, the Society lost its creativity and became simply an upper-class club.

In the 18th century, local groups of amateur scientists in Scotland and provincial English cities exhibited more creative vitality than the Royal Society. Yet this also proved to be a passing phenomenon, limited to the generation of activists who had initiated these societies. When the founders were succeeded by the younger men they themselves had elected to membership, the vigor of the initial period was no longer in evidence. In the end, rejuvenation of the learned society was accomplished only through open membership. The British Association for the Advancement of Science was formed on this principle in the 1830s, one of many peripatetic organizations formed in a period of conservative reform following incessant wars and working-class disaffection. It not only incorporated the scientific curiosity of the top of the middle class (mostly clergy, academics, physicians, and some aristocrats) but pioneered discussion meetings (which the Royal Society had never had), government lobbying, and research grants (Thackray, 1983). The Royal Society was outflanked.

The Académie des Sciences was less a learned society than a national research institute, fully integrated into the French bureaucratic system. It was not a circle of like-minded associates but an institution that reigned over all groups of scientists in the land, threatening at times to turn into an instrument of oppression. The profusion of provincial academies that appeared in imitation of the Paris Académie during the 18th century were dominated by a conservative atmosphere that reflected their founding by members of the privileged classes—local aristocrats, procurators, and priests. Subsequent history indicates that thenceforth emergent groups of scientists were to seek the protection and patronage of power. Binding themselves closely to the established order, they took as their model not the Royal Society (which I will call the society type) but rather the Académie des Sciences (which I will call the academy type).

Societies of the academy type also flourished in Germany. Having despaired of the university as a place where learning could be advanced and promoted, Leibnitz devoted the last half of his life to the formation of this type of society. The result was the Berlin Academy, founded in 1711. Rulers in other principalities followed suit, each desirous of having one of these regal ornaments and advisory organs of his own. Generally speaking, these academies were modeled on Paris, in imitation of the court of Louis, although the rulers of these small states could not pay the high salaries that attracted first-class foreign scholars to Paris. The ornamental presidents of these academies were given aristocratic titles or selected from among persons with aristocratic family backgrounds. (This idea that such bodies should be headed by a titled personage also seems to have been firmly ingrained in the Japanese mind, for when the Physics and Chemistry Research Institute was founded in 1917 its first chairmen were chosen from among the aristocracy.)

The French conception of the learned society as a national institute of science is one frequently found among revolutionary regimes. Revolutions are almost always preceded by criticism of the universities. The French Encyclopedist ideologues vociferously attacked the *ancien régime* at the University of Paris. The intelligentsia of Imperial Russia had similar antipathies. The abuses of the university—and overseas student cliques—in pre-1911 China were so outrageous that they became the source of much heated controversy. Whether or not students really are the vanguard of revolution, a preparatory outbreak of attacks on the university would seem to be written into the historical scenario.

The university reform programs of successful revolutionary regimes would also seem to have several common features. To combat the inefficiency and stagnation of the university as a research institution, these programs typically call for the separation of education and research and for the creation of one or more highly efficient national research institutes on the one hand and of a system of technical schools on the other. Since revolutionary regimes ordinarily seek to centralize power, they tend to build large, state-run, central research centers.

Thus the Moscow Academy of Sciences was modeled on the French Académie des Sciences and, more immediately, on the Kaiser Wilhelm Society of Germany (now the Max Planck Institute), and the Chinese Academy of Sciences was constructed on the Soviet model. Today there are many academies of science modeled on the Moscow Academy throughout Eastern Europe and the Socialist bloc. Operating as research and administrative organizations, they exist apart from—and superior to—those institutions whose function is educational, that is, the universities.

These academies would seem to be highly rational and efficient research organizations, the very embodiment of the Baconian ideal, in which project planning can be adequately monitored and wasteful duplication of research avoided. Yet this type of structure easily becomes a mechanism of oppression, snuffing out the first expressions of new modes of learning. With the reputation and authority of a National Academy, almost any "new wave" can be held in check.

To cite one example, the members of the French Académie des Sciences were highly accomplished, established scientists who, among their many other activities, awarded a variety of prizes to encourage scientific inquiry. Inasmuch as government funds were being used to stimulate scholarship, this was obviously a good thing. Nevertheless, the judges were all established "greats," old men who could not really appreciate work that was not an extension of their own. In short, the prizes went only to the work of normal science. The applicants themselves knew who sat on the judging committee and submitted only work they thought would impress them. Murmurings of dissatisfaction at the high-handedness of the academy members were constant, and the occasional scandal even developed.

Voluntary Associations

Rebellion against learned societies based on the academy model did not become widespread until the 19th century. When it came, the revolt centered around the principle of open membership and the desire for academic societies that would serve the needs of more specialized constituencies. In restricting their membership, the old-style learned societies had become closed, honorary organizations dominated by an elitist atmosphere. It was in opposition to these conditions and as the locus of a nascent movement to disseminate science and elevate the social status of scientists that democratic organizations, open to anyone, came into being.

The following are among the main societies that emerged from this movement:

- 1815 Société helvétique des sciences naturelles
- 1822 Gesellschaft deutscher Naturforscher und Ärzt (GDNA)
- 1831 British Association for the Advancement of Science (BAAS)
- 1833 Congrès scientifique de France
- 1848 American Association for the Advancement of Science (AAAS)
- 1872 Association français pour l'avancement des sciences (AFAS)
- 1907 Società italiana per il progresso delle scienze (SIPS)

The originator of the modern academic association, Lorenz Oken (1779–1851), was a member of the Societé helvetique. It was he who hit upon the idea of forming a pan-Germanic academic society. The Gesellschaft deutscher Naturforscher und Ärzt (GDNA) which he organized was so unusual that it was suspected of being a front for a revolutionary political movement, and there were several attempts to suppress it. At the first meeting, fear of the authorities kept many of those who actually participated from registering. Despite such problems, however, this annual meeting, held each year in a different German city, generated an extraordinary amount of enthusiasm within the scientific world.

The post-ideological scientist of today may think that the government interference was simply bad bureaucratic judgment. But in those days the ideological significance of science was more obvious than it is today. The government had ample reason for keeping its eye on this association of scholars.

Under the system of small states into which Germany was still divided, the interests of a rising bourgeoisie intent on certain English-style liberties and German unification came into conflict with the local administrators and nobility who sought to preserve the old order. It was a time similar to the second decade of the Meiji era (1877–1887), when the Japanese government kept a careful eye on the Movement for Liberty and Popular Rights. Against the logic of a territorially divided order, the logic of Germanophone science pressed ideologically toward a thorough-going universalism and full liberty, and many of the new scientists who advocated it in fact came from bourgeois households in the lower middle stratum of society. If they were subject to surveillance, it was because they shared bourgeois aspirations for liberty and unification. In order to avoid intervention by the authorities, Oken insisted that papers presented at meetings of the society limit their discussions to the purely scientific. But there was a real possibility that someone would launch into a discussion of German unification and liberal democracy at one of their gatherings.

The 19th-century German university was to become a remarkable research institution (see Chapter 5). Yet the vitality of 19th-century German science is inconceivable without the GDNA and other modern academic associations. It was because of their efforts that scholarly traditions invaded the German university in search of professional working conditions for researchers and a mechanism for group replenishment. In this as in other ways, the GDNA clearly distinguished itself from learned societies of the older type.

A group of men who see each other all the time is not the most likely setting for an outpouring of new ideas. Fresh encounters between people who seldom meet are more important. Such contacts are particularly decisive during the formative period when a paradigm is taking shape. The Royal Society was a London organization; the Académie des Sciences, a Parisian one. Scholars who did not reside in these cities were, in principle, not admitted as regular members. Meetings of the old-style learned societies were gatherings of the same faces—unlikely places for tension and excitement or a fresh challenge. (Some sociologists argue that researchers who always work together in the same laboratory are liable to lose their drive and thus perform their work merely from force of habit, though the same arrangement seems to be effective in expediting and developing the routine research involved in normal science projects.) In the light of these observations we can fully appreciate the psychological significance of an annual general meeting of the GDNA.

The GDNA was, for all practical purposes, organized on the basis of open membership, as the publication of a single scientific article made one eligible. This association grew rapidly in the

German states and also served as a model for foreign members who founded similar organizations in their own countries.

The emergence of this type of assembly was not unrelated to the growth of railway transport in the 19th century. The trains made it possible for the BAAS to draw members from all over Britain and avoid automatic domination by London. The activities of local academies and associations peaked during the 1870s. Thereafter, decentralization proceeded rapidly. Pan-European, international associations were formed, and with the spread of air travel after World War I, academic associations in North America experienced a wave of internationalization. Eventually, after World War II, this wave would beat upon Japanese shores.

The older learned societies had little to do with the remarkable advance of specialization that occurred in the sciences during the 19th century. In fact, they were often centers of resistance to this trend. The new academic associations that emerged to compete with them at first divided into sections and carried on specialized activities. Close on the heels of this development, associations of specialists were formed. In England, for instance, the Geological Society (1807) and the Astronomical Society (1820) were the first in a succession of groups to exist despite the opposition of the Royal Society.

As an adornment of the State, the old-style academies had received the generous patronage of government. But governments could not commit themselves to automatically subsidize the new, neither-fish-nor-fowl modes of learning and the many specialized fields that were rapidly proliferating. In this situation, societies for the promotion of science came into being, societies whose goal was not only the advancement of scientific knowledge but clearly the promotion of the scientist as well. The BAAS extolled the contribution of science to the state and unveiled a campaign to sell the scientist to the general public as a servant of the nation.

Chapter 5

A Century of Professional Specialization

Academic standards can be said to exist wherever two or more scholars are engaged in an ongoing conversation about issues and questions that pertain to a common specialty. One of the functions of academies and learned societies is to help preserve these standards. Academic traditions, however, involve more than groups of scholars and the common standards they would uphold. In the long run, one may speak of an academic tradition only when a younger generation of scholars is able to make a living in the pursuits of their predecessors—that is, only when more or less permanent arrangements have been made for education and group reproduction. The emergence of a group of scholars in support of a paradigm may result in the maintenance of academic standards for a time, but unless the group complements its horizontal ties by gaining access to the university and by building a vertical tradition, these paradigmatic standards are likely to be relatively short-lived. No matter how lively the academy's discussions, there is always a danger that they will die out after one generation.

The emergence of the entity we know as modern science is almost universally associated with the 17th century. Yet science was never more than a movement among one segment of the intellectual community during this period. It was not until the 19th century that the institutional conditions which were to underwrite its full independence came into existence. These were the years in which science became professionalized and the scientist acquired the clearly defined and well-established position in society that he enjoys today. In this step to maturity the university was to play a central role.

The Establishment of German Academism

Today unanswered questions are brought to the university, but if an 18th-century European had a problem, chances are he looked toward the learned society or academy for a solution. In other words, the popular notion of the university as the seat of boundless knowledge, ready to supply an answer to every inquiry, is not very old. It may in fact be traced to the prestigious heights attained by the German university in the 19th century—to the rise of what is known as German academism.[1]

Living as we do in a world dominated by superpowers who are, at the same time, the chief protagonists of "big science," we are apt to think that science and learning develop only under conditions of national eminence. During the rise of the German university (a period which runs from the last years of the 18th into the first half of the 19th century), however, the German states were not yet united. Ransacked by Napoleon and still governed by a host of territorial rulers, Germany was without effective power as a nation. Why then did the German university become the most vital center of academic research in Europe during this period?

Attempts to see this development as a consequence of industrialization are pointless, for the growth of academic science in the German university seems to have preceded the industrial revolution. During the period in question, Germany continued to lag behind France, not to mention England, and the German bourgeoisie remained immature. Friederich Paulsen has even cited "retarded social development" to explain why the best German brains entered the academic world rather than go into politics or industry (Paulsen, 1906). Whether or not a detailed sociological analysis would uphold this interpretation, it is true that the academic world did offer almost the only course of advancement for an ambitious German youth in the 19th century.

Nor can the prominence of German higher education be explained by calling it a new system specifically devised by political authorities for the promotion of scientific growth. Göttingen, the first of the modern German universities (founded in 1732), began

[1] As used in Japan, the word *akademizumu* (academism) has peculiar flavor about it and practically speaking it can be thought of as a Japanese neologism.

quite unpretentiously. Indeed, all that can be said to explain its distinction is that the small size of both the state and the university may have made adaptations to new environmental conditions easier than elsewhere. In the area of governmental policy and planning, the institutions created by the French Revolution were far and away more original and forward-looking, and Germany strove hard to imitate them, as well as those of the English, until well into the 1830s.

University Autonomy

The achievements of the German university have also been attributed to its tradition of autonomy and strong sense of academic freedom. Here one would seem to be on at least somewhat firmer ground, for, after nearly two centuries of sometimes precarious existence, the first half of the 19th century did witness the triumph of the autonomous university, at least as the ideal. But the concept of university autonomy is essentially a defensive one. While the university seeks institutional guarantees for academic freedom, it does not actively contribute to the cultivation and advancement of new scholarly tradition. In other words, it seeks to protect intellectual activity from the pressures and constraints of extra-academic values, leaving the process of paradigm creation and selection open to determination by scholars themselves.

Thus the triumph of this ideal did not necessarily mean that German university men were to live and work in a free atmosphere. Through structures of self-government persons within the university manage their own affairs. While these structures guard against external infringements of liberty, they can also be used to shelter the privileged guild-controlled scholarship that has already established itself and to suppress the development of new modes of learning. They tend to work to the disadvantage of those seeking to get in. Certainly the new group of liberal scientists who emerged in the German states during the 1840s saw these structures as antagonistic to scientific advance. Rules and regulations governing faculty membership were also restrictive, making it difficult for atheists, Jews, and socialists to obtain university positions. All these things prompted Wilhelm Humbolt, founder and first president of Berlin University, to hope that the views of an enlightened government might bring a breath of fresh air to university life.

Political Decentralization and the University

If the several factors considered above fail to account for the rise of the 19th-century German university, what other possible explanations are there? Sociologist Joseph Ben-David has sought the answer in Germany's decentralized sociopolitical system.

In science the early years of the 19th century belonged to France. The new institutions created under the revolution had begun to demonstrate their efficacy, and Frenchmen prided themselves on the rationality of a system which after Napoleon clearly reflected the French penchant for centralization around the Parisian axis. Under these non-competitive conditions the new institutions began to age quickly, though the educational system remained almost unchanged until the First World War and even thereafter. (The École Pratique des Hautes Études, founded in 1865, was the sole exception. Here, outside the mainstream of the French educational system, new fields of scholarship were developed in areas where traditional disciplines impinged on one another.) Regrettably, the human products of this system continued until recently to have absolute confidence in the superiority of the centralized rationality and efficiency of their own institutions. This is the kind of thinking one expects in countries dominated by a bureaucratic elite. We are reminded that France had been the most enthusiastic imitator of the Chinese civil service examination system ever since the Jesuits introduced it to Europe.

Meanwhile, scholars in Germany maintained some freedom in their choice of institution amidst the local separatism of the territorial princes and the balance of power that existed among them. Sustained by a competitive desire to improve their own domains, rulers vied with one another in making their institutions more attractive, in improving the treatment accorded faculty, and in luring away outstanding teachers from other states. In order to raise the prestige of their territories, they sought out famous scholars much as local government officials today work to attract industry. Estimates were made of the number of students these scholars would attract and the amount of money the projected number would spend in their cities. This increased the competition between individual scholars who wanted higher status and better working conditions, and encouraged students to travel in search of better teachers.

In terms of a systematic policy for the encouragement of research, however, France remained superior. The French government was even more conscious of the value and utility of science than the German rulers, and they awarded prizes and positions in the university or national academy to outstanding scholars. But these honors and awards went to established scholars who had already made a name for themselves, not to young men who might have used them to further their research.

Since one could not secure a university post in 19th-century France until he was in his forties or fifties, young men were unable to make a living as scientific researchers during the most creative period in their lives.

In the early 19th century, scientific research was largely an amateur enterprise throughout Europe. The majority of those engaged in it were doctors, pastors, lawyers, or men in some other profession. They did research as a hobby and bore the cost themselves. Yet the absence of the conditions that make professional practice possible inhibited specialization. When the aspiring scholar does not know which of a very few positions will open, or when, he must maintain a range of research interests and activities that is wide enough to enable him to fill any position that might come along. He must be a jack of several trades. Indeed, where this mechanism is operative, the narrow specialist is treated with disdain. The French term *savant* (scholar) referred to a generalist, to a versatile man with extensive knowledge in many areas.

It was in Germany that scientific research first became a profession. This development had its origins in negotiations between outstanding individual scholars and the universities that bid for their services: it was not the result of a unified national plan. "Name" scholars were simply in a strong enough position to negotiate working conditions with their employers. Their contracts might stipulate, for instance, the right to give lectures in a new field. And once a promising new discipline was publicly recognized at one university, other institutions would follow suit.

Moreover, the *Privatdozent* system made it possible for young scholars to live directly on the honoraria paid by students while giving experimental lectures in new areas even before a regular post was within their reach. Specialization continued to make advances, and as specialized scholarship became professionally viable, amateurs were no longer to be seen at the forefront of scientific research.

If national academies and research institutes reflect a centralizing impulse, the university is decentralizing. The idea of gathering the top scholars of the day once captivated enlightened despots and later proved to be quite compatible with the rationality of revolutionary regimes and their concern with efficiency in research. Thus major research institutes were built first in Paris and, later, in Moscow and Peking. The university, on the other hand, being a place for the teaching of students, finds the hiring of several specialists in the same field both unnecessary and extravagant. The result is a dispersal of specialists in any given field. This may afford an occasion for fruitful intercourse among scholars in different fields. At the same time, it creates a need for national academic associations that bring together all scholars in the same field. As we have seen, this kind of association was first formed in Germany and eventually spread throughout the modern world, replacing the old-style academy whose life membership system and restrictive attitudes toward recruitment had led to stagnation.

Unlike normal science, the creation and development of scholarly paradigms occurs at times and places so indeterminate as to exceed the bounds of human predictibility. Even if one fashions a master plan that carefully incorporates the most rational and enlightened ideas, it is simply not possible to map out an enduring science policy. Of course plans can be drawn up, but like those of the French revolutionary regime, they will soon be superannuated. That the decentralized German system outstripped the French system in the advancement of learning and professional specialization is one of the ironies of history. At the same time, it suggests that we must always be ready to rework our plans whenever new circumstances arise.

But let us look at the rise of the German university from another angle. Decentralization may have been a significant factor, but it was certainly not a new one. The competitive conditions it provided had long been in existence. Local universities were vying with each other in the 17th century, and, as we have seen, 18th-century Germany not only produced more journals than any other nation but abounded in local academies as well. The failure of these old-style academies and learned societies to keep pace (either in training successors or in specialization) and the growing concentration of scholarship in the university that resulted may help account for the sudden upsurge of German scholarship in the 19th century. Beyond this general consideration, however, one must note the peculiar influence of the Napoleonic wars.

Universities in 18th-century Germany were so dead that in some instances plans were even made for their abolition. As Napoleon overran the country, certain features of the old university order were destroyed. From the midst of post-war confusion came Berlin and other new institutions that were to serve as modern models. These institutions challenged the old universities and drove them to reform. It was in this way that the wine of new science and learning was poured even into the old wineskins of the German university.

The New Science and the Educational Establishment: The English Case

In England, the conflict between the supporters of the newer-type universities and the defenders of the older colleges became the focal point of a debate over higher educational reform that continued throughout the first half of the 19th century. The college-centered individual guidance that lay at the heart of Oxbridge education might have been good enough for providing the sons of the wellborn with a genteel education, but many practitioners of the modern sciences argued that instruction in the new specialized disciplines required a return to the lecture system of the medieval university.

The founder of modern geology, Charles Lyell, charged that the colleges were breeding grounds for a strong guild consciousness that left no room for the introduction of new learning, and havens for a species of men for whom personal considerations were more important than scholarship (Tillyard, 1913). Yet hereditary estates and other endowments gave the colleges a prosperity and a power that stymied attempts at internal reform. Despairing of this situation, Lyell placed his hopes in the work of the Royal Commission. Meanwhile, the philosopher William Whewell (1794–1866) championed the education of the colleges and opposed the work of the Commission as a violation of university autonomy.

Whewell divided the English curriculum as it stood during the first half of the 19th century into the "permanent" and the "evolving." The former consisted of the fully mature disciplines for which there were standard, printed texts—fields like mathematics and the classics. These were to be learned with a tutor in the colleges, where the emphasis was less on their development as scholarly fields than on their significance as training for the mind. The natural sciences, on the other hand, Whewell placed in the evolving category. Since these studies were as yet unfinished, the time had not yet come when they could be put into a course of study and taught from a textbook. These developing disciplines were properly taught from the lecture podiums of the universities.

If university lectures were now conceived as vehicles for communicating the issues and opinions being discussed in intellectual and academic circles, the situation was, after all, quite different from that of the Middle Ages when there was no printing, and handwritten manuscripts recorded from lectures were the chief medium of scholarly communication. To conservatives, however, the lecture-centered German-style university in which instructors bandied about scholarship still in the making inevitably meant criticism of the established order. Should this kind of system be adopted in England, established religion would become a particularly vulnerable target. An educational institution that would poison youthful minds with such criticism was surely unacceptable. Had not Lyell's lectures on geology already led to criticism of the Biblical account of creation? For his part Lyell wearied of conservative attacks and discontinued his lectures after two years.

After nearly half a century of bitter debate between the promoters of the new sciences who increasingly sought government support (proponents of the German-style university) and defenders of college autonomy (those who sought to maintain the tutor system unchanged) produced no substantative conclusions, a government commission moved in with its own study, and there were signs that Francis Bacon's advice to the universities of his day to "seek out neglected fields and publicly appoint scholars to work on them" was finally going to be taken—more than two hundred years after it had first been given.

If academic freedom is the liberty of scholars to construct and select paradigms as their interests and concerns dictate, there is no reason why the university is necessary to its enjoyment. Members of the Royal Society had much more latitude in these matters than 17th-century German university men. The story of Spinoza declining a call to a professorship at Heidelberg in order to insure his independence of thought, despite initial assurances that he would have full philosophical liberty, is well known. One even feels that the "Dutch scholars" of Edo Japan had more intellectual freedom than the university professors of the Meiji period, who were also civil servants. "Academic freedom" is then not to be exclusively identified with the university; it is a part of the corpus of civil rights.

The contemporary notion that university autonomy is the one and only fortress of freedom may be indicative of the strategic significance the university has had under a variety of historical conditions as the last bastion of freedom of inquiry, but it is presumptuous to think of this freedom as a special right. More seriously, preoccupation with university autonomy may cause us to lose sight of the real threat to "academic freedom" today—the surging tide of bureaucratism and power-elitism that has swept over scientific research.

Scholarship in the University: Textbook Style

When the university replaced the academy as the center of scholarly activity, the shift was accompanied by appreciable changes in the conduct of research and in perceptions of what scholarly activity ought to be—changes which our general understanding of the response of institutional structures to new intrusions and situations would lead us to expect. While these changes in the character of science attendant on its removal from the 17th- and 18th-century academies to the 19th-century universities promise to be one of the major themes of future studies in the history of science, I would like to record some preliminary observations on several crucial aspects of this transformation.

In the academy, the scholar could discuss whatever subject he wished. As a member of a society of fellows, he and his colleagues met on common ground. The university, however, is built around two distinct strata. The scholarly activity that goes on there presupposes not only scholars but students. This is a circumstance unknown to learning in the academy. While scholars are caught up in the discussions of the academy they feel little need of students. Busy producing and developing new ideas, they have no time to consolidate the results for any purpose except communication to peers. But when a scholar is confronted with a university teaching assignment, he must organize his thoughts so that they are comprehensible to the uninitiated. The need to develop a course of study provides him with an opportunity to reflect on the research he has been doing, and encourages the development of a unified set of technical terms, a system of classification, and articulation of methodology. When they don professorial garb, even researchers working at the frontiers of their fields are obliged to explain their work in terms of basic principles, gathering up and sorting out until they have put together a rational compendium of knowledge. In short, the university science lecture platform regularly turns out scholarly work cast in a broad conceptual framework and written in the style of a textbook.

The learned societies and clubs of the 17th century were without systematic structures for nurturing successors. When a group or school that had formed around a scholarly paradigm had resolved most of the problems it posed and ceased to generate new ones, its work no longer engaged the imagination and the group lost its vitality. As the remains and relics of its scholarly activity were enfolded in encyclopedic dictionaries, manuals, and handbooks, the scholarly tradition became defunct. These encyclopedias (for general use) and manuals (for the specialist) were quite sufficient in the 18th century, when scientific education remained a largely unsystematic affair, and amateur scientists and high-level artisans learned science on their own. But as science became normal science and normal science grew more and more complex, looking at manuals and encyclopedias was no longer enough. A certain amount of organized, sustained, step-by-step training over a period of years came to be necessary. Conversely, when scholarly work showed this kind of maturity, it was ready to be incorporated into the educational curriculum and taught in schools.

Under these conditions, solution of the major problems generated by the paradigm is followed not by death but by stabilization of the tradition and its reformulation in textbook form. Since the value of a textbook lies in efficient presentation of a finished product for ready mastery, it omits the dead ends, the unessential aspects, the trials and errors of the development process, and prefers a logically ordered step-by-step explanation of the paradigm to a descriptive, chronological account of its development. The name of the paradigm maker and founding father disappears: Newtonian mechanics and Cartesian mathematics become mechanics and analytical geometry. It is this kind of coherent knowledge and these kinds of settled techniques brought together in textbook form that are best suited to the lecture format and can be most easily taught in the university in a limited period of time.

If scholarship in the paperless medieval university was synonymous with whatever could be presented in lecture form, the advent of the modern university brought the popular notion that scholarship was something that could be outlined and articulated in general concepts and written down between the covers of a text. Today's academic departments (physics, chemistry, geology, etc.) are, in the main, products of the 19th century, but the faculty and departmental composition of the world's universities and the professional groups predicated in them are also reinforced by and through textbooks. Thus, if one can speak of a group of men who embraced the Confucian classics as Confucian scholars, one can also say that modern scholarly groups—at least the larger and better-established ones—are built around common textbooks. It is true that such texts lack the authority of classical canons, venerated by Confucians for their antiquity. Indeed the modern textbook is supposed to include all the significant aspects of relevant scholarly activity to date, and needs to be constantly revised to take account of the latest achievements.

The development of textbooks itself contains within it the seeds of a division between education and research. Textbooks and courses of study are compartments designed for students rather than for those at the cutting edge of research, and often they even inhibit the generation of scientific revolutions. Still, it is only in what Whewell called the "permanent disciplines" that scholarly paradigms can be cast successfully in textbook form. Paradigms in such fields as differential and analytical calculus continue to be regularly studied even though they no longer generate new problems, because their utility as tools for other disciplines has been recognized and they have become an unchanging part of the basic curriculum. The study of languages and mathematical techniques falls into this category. These subjects are normally taught as required courses and treated as skills which call for the same kind of repetitive practice it takes to learn to drive a car.

Experimentation and Its Scope

What the medieval university taught, it taught in the form of lectures, discussion, and debates. Not only law and theology, but even medicine was taught through commentaries on classical texts. Those things that could not be communicated through lectures—the practical and performing arts and those empirical studies that required observation and experience—were not included in the university curriculum.

In this sense mathematics occupied a peculiar position among the disciplines as we know them today. Of course mathematics had been part of the classical liberal arts curriculum, but it owes its position in the pre-modern university to the fact that it was not a positive science. Even in early 19th-century Germany, natural philosophy normally meant eloquent lectures designed to elicit the awe and admiration of students, not empirical science based on positive research. It was the last hurrah of science as pedagogical, rhetorical learning.

Before this period those who wished to conduct experiments simply remodeled their kitchens and proceeded. Neither universities nor governments provided special financial assistance or accommodations for experimental apparatus. Thus in 1825, when Justus von Liebig returned to Germany from study at the École Polytechnique and set up a small laboratory for students at Giessen University, a radically new step had been taken in the history of higher education. Scholars had built experimental labs for themselves as early as the 18th century, but these were temporary affairs that were dismantled after the research for which they had been created was completed. Liebig, on the other hand, envisaged his laboratory as a place where students could be taught the systematic approach and methods of analysis he had learned under Gay-Lussac. Since a new generation of users arrived every year, laboratories for students were permanent. It was not long before the idea that laboratories were indispensable to scientific research became generally accepted in the university world. From the 1840s through the 1860s, a dispersion of chemists trained in Liebig's laboratory spread the gospel to universities throughout the world, to industry, and to the field of biology. Physics, however, responded more slowly. University laboratories began to be built for the study of physics in the 1860s and 1870s, but physics was to be forty years behind chemistry in becoming a full-fledged laboratory science (Cardwell, 1963). The discipline continued to be dominated by Newtonian mechanics. "Industrial sciences" such as the study of electricity and calorifics were initially the work of chemists.

During the course of the 19th century, the image of modern science as something done in red brick buildings by scholars in laboratory attire became well established. As laboratory sciences such as chemistry and biology asserted themselves as university subjects, university men came to look upon field-oriented sciences involving the collection of specimens, like natural history, as unscholarly, amateurish pastimes. Charles Darwin's theory of evolution did not emerge from within the university. Engineering was also shut out of the German universities and those they influenced, because it was felt that the practical training so vital to education in these areas could be carried out quite satisfactorily by apprenticing people on the job and was not something in which the university should get involved. Those subjects that could be taught and researched within the scope of university facilities were considered paramount. Nationwide geological surveys and big science projects that exceeded the scope of the university were politely shunned by university men, who tended to view them as low-grade, unscholarly endeavors.

The Home of Normal Science

The modal subject taught at the university is a normal science, a mode of learning grounded in a publicly sanctioned paradigm. The construction of paradigms—how to make a scientific revolution—cannot be taught in ordinary ways. How significant is what is taught at the university for someone who wishes to become a writer? Today drama, music, painting, sculpture, architecture, and other arts are part of the university curriculum, but this is chiefly because a university degree has become a status symbol, and higher education has become a buyer's market. By gaining access to this degree, practicing artists hope to elevate the social status of their profession, even though the university is more suited to training critics than artists. The basic skills required to execute a piece of music can be taught, but more diffuse skills—how to make money, how to be a successful politician, and so on—lack a paradigm firm enough to serve as the basis for the development of a tradition of learning.

The tempo of classical university education is also incompatible with fields in which the pace of progress is extremely rapid. In contemporary *avant-garde* mathematics, for instance, a new paradigm appears, a group or school supporting it forms, and most of the new problems engendered by the paradigm are solved with lightning speed—whereupon another new paradigm appears from a different quarter. Given this kind of situation, what the student learns from long years at the university and in graduate school studying received paradigms and developing his problem-solving capacities is already out of date by the time he begins to think about doing front-line work. Indeed, this kind of training may well have the effect of repressing the individual's capacity to generate new ideas.

Scholarship as a Profession

Thus far we have looked at the incorporation of science into the university chiefly in terms of its consequences for the character of scholarship itself. But the university, unlike the academy or learned society, is not an institution for scholarly research alone.

Ever since the Middle Ages, it has been a guild for those members of the intellectual professions who have made it their home. For modern states and industrialists, it has been the basis of a recruitment system, while for those seeking to get in, it has been a gateway to success. It has been a place where all kinds of secular interests compete. Let us look then at this center of clashing interests from the student side of the fence, for students must carry a major share of the burden if the university is to fulfill its special function as a group-replenishment mechanism.

Isolated amateurs or members of a learned society characteristically have other means of support, but for the university student, study is directly related to a future means of livelihood. Thus learning in the university is shaped by occupational requirements in a way that the scholarly activity of a learned society theoretically is not. The traditionally preeminent faculties of theology, law, and medicine were built upon links with the old guilds and the situation had not changed much by the 19th century. Students developed their interests in new fields of learning such as the sciences or history while studying theology, for instance, and preparing to work for the church or a related organization. Law and medical students might develop similar interests, but there were few opportunities to practice science professionally, and few students were inclined to do so. In short, the new disciplines were hobbies for upper-class intellectuals. They were not qualifications for employment.

Thus the first scientists were government officials, self-employed professionals such as doctors or private school masters, and well-heeled amateurs. Eventually, however, some found that scientific work absorbed more and more of their time and interest. They began to think of themselves primarily as scientists and looked forward to the time when they would be more than hangers-on, when science would be an independent learned profession similar to law and medicine. (In China everyone sought to follow the bureaucratic route to success, and independent professions did not develop, much less modern science.)

That research-oriented scholarship became a part of the university owes much to the rise of faculties of philosophy in the German universities. In the medieval university, philosophy had been part of a liberal arts curriculum that ranked below the specialized faculties of theology, law, and medicine, but it eventually rose to equal that of the other three faculties.

Oxford and Cambridge had been closed to all but aristocrats and members of the propertied classes because of the large financial burden they placed on parents, but in Germany governmental contributions amounting to eight times what was received from student tuition and fees considerably reduced the cost of higher education and opened the university to the middle class. The philosophy faculties offered particular advantages to middle-class families. In the traditional fields of law and medicine a young graduate usually had to depend on his family for several years before he could get himself established. With the rise of the bourgeoisie, however, the teaching profession (long a clerical monopoly) began to open up, and many secondary school positions (in the German *gymnasium*, the French *lycée*) became available to graduates of the philosophy faculties. Since these jobs could be had immediately upon graduation, the financial burden of professional education in this case did not go beyond what parents who were lower government officials or elementary school teachers could bear. Thus the new group of scientists was largely composed of middle-class talent. If one had the ability he might eventually rise to become a university professor and rub elbows with the upper classes. In the 17th century science was an upper-class gentleman's accomplishment, but during the 19th century the scientist's social stratum fell even as his numbers increased.

Even in the United States, where the natural sciences and technology have long enjoyed a prominent position, scions of the professionals in the eastern states have preferred to study law or medicine. It is the middle-class families of the middle and western parts of the country that have produced the scientists (Goodrich, 1962). Here, as elsewhere, scientists have tended to come from strata that have enough money to send their children to college but are unwilling or unable to support them while they establish themselves afterwards.

With the advent of science as a profession and a viable way to make a living, scrambling up the ladder with bare hands was no longer the only route to success. Research became a more rigorous business. Amateur researchers were deposed and replaced chiefly by graduates of philosophy faculties. This was science for passing through the portals of the university, science for conquering the obstacle course that led to a higher degree, science for acquiring the professional qualifications that had heretofore been the preserve of a social elite. No longer would the scientist be exclusively identified in the popular mind with a special breed holed up somewhere in a laboratory. Through the university, it had become possible to distinguish the professional scientist from the amateur, to nurture and reproduce him, and to establish thereby the formal conditions for a new profession.

Learning for Qualifications

Once word gets out that science is a rising stock, students pour into its classrooms. This leads both to an increase in the number of scientists being produced and to a standardization of quality. Unlike the limited number of gifted students who pursue science because they find it personally compelling, those who select science because it offers a promising future want an academic degree as a passport to employment. The study of science becomes, in short, a qualification.

The granting of qualifications is perhaps the major function of the university. Accordingly, learning for qualification reigns supreme. The law and medical faculties remain dominant within the university because they qualify students for prestigious professions. Those disciplines that are readily testable by examination come to be regarded as more teachable than others. Fluency in foreign languages is considered difficult to teach, even though instruction in grammar proceeds smoothly, because spoken fluency is difficult to measure in examinations.

That university examinations play a significant role in shaping academic traditions cannot be denied. To cite one example, owing to the ease with which mathematics could be tested, the mathematics tripos at Cambridge University came to be regarded as the highest measure of intellectual ability. Under the influence of this examination, 19th- and early 20th-century Cambridge-educated scientists proved adept at handling mathematical problems but—with the exception of Clerk Maxwell—lacked the kind of insight into physical nature that the construction of scientific paradigms requires (Crowther, 1952).

A professor may give impressive papers at scholarly gatherings and deliver well-polished lectures, but unless he gives exams, and unless these exams are related to the acquisition of qualifications, he will not have much of a student following. The fact that newly developed fields of learning are not immediately accompanied by a definitive set of exams and qualifications is one of the reasons they have difficulty gaining student acceptance.

A system of qualifications acquired through written examinations was first introduced into Europe by Jesuit missionaries, who had been impressed by the efficacy of the Chinese civil service examinations. In the latter half of the 19th century, written exams enjoyed an unprecedented boom and came to be widely used both within the educational system and in the selection of government officials. But as I have argued in Chapter 3, this kind of system is by no means suited to exploring and developing the frontiers of knowledge. It has become a part of the university simply because it is indispensable to the professionalization of science.

Scholarship and Professional Consciousness

With the professionalization of science, scientists become aware of themselves as professionals and wish to be regarded as distinct from mere amateurs.

At meetings of learned societies or gatherings of amateur enthusiasts, whatever is of interest at the moment can be discussed, be it science or literature, art or politics. In fact, records of such societies reveal more than one non-scientific discussion. The Royal Society was the scene of many distinguished research reports, but it also listened with apparent willingness to a forthright account of the sighting of a monster in a nearby village. At an institution for training professionals, like the university, however, amateurish, pre-paradigmatic knowledge was looked down upon as unscholarly, an attitude which manifests itself in an atmosphere that would prefer to dismiss all talk of politics as "barber-shop conversation."

This attitude almost inevitably finds active expression in the deliberate attempts of professional scientists to build restraining walls designed to keep amateurs away. Institutionally these are the walls of the university. Within these walls the concept of "science for its own sake" makes its appearance as the de facto claim for allegiance owed only to the values of one's professional groups.

Academic degrees also exist to distinguish scholars from others. Studies for which academic degrees cannot be conferred have no connection with the university and are not recognized as proper academic disciplines. For scholars engaged in scientific research, access to expensive instruments for measurement and observation (e.g. giant telescopes) is a particularly effective way of drawing a line between themselves and dilettantes. Until the 19th century, wealthy amateurs could still buy materials for constructing experimental apparatus in their own homes, and no one dreamed that universities might one day monopolize the tools of research. But in the age of big science in which we now live, access to and ability to use expensive experimental facilities and research methods have become an important way of decisively separating professionals from amateurs.

The use of technical terms and concepts even when everyday language would do as well is yet another device by which professionals seek to set themselves apart from others. Of course logical precision cannot be preserved without the use of clearly defined technical terms, but one often notices a deliberate tendency for scholars to use specialized language in order to make their work sound authoritative. The technical terms coined in Japan in the early Meiji period (particularly in philosophy) were the product of nascent academic professionals who, in the course of attempting to assert their claim to a distinct social identity, deliberately invented artificial words to translate concepts that had originally been expressed in quite ordinary English or German. As a result, specialists were no longer able to talk with amateurs, and higher learning was deprived of potential resonance with popular culture; but this in itself enhanced the specialist's authority.

Of course the world of the professional has stringent demands of its own. When an amateur tires of doing research he can simply quit, but once he becomes involved professionally things are not quite so simple. Even after intellectual creativity is gone the professional scholar-researcher must carry on—a tragic situation that is not only painful for the individual concerned but creates problems for others as well. Moreover, a professional specialist does not have the amateur's freedom to follow changes in his own interests and flit about like a butterfly outside his special field. Having hung out his shingle as an expert in a given area and a teacher of a particular subject, he cannot simply ignore the expectations of his students, colleagues, and employers.

Again, an amateur can, if he likes, work on problems that have already been resolved by others or otherwise have no significance for the advancement of learning, but for a scholar-researcher to present received knowledge as his own work is a cardinal violation of professional ethics. Thus he must begin his research with a thorough survey of what has previously been done in the field, cite in footnotes other studies he has used, and generally strive to make a clear distinction between his own contribution and the work of others. It is perhaps not coincidental that the use of footnotes became a fixture of academic writing style during the same latter half of the 19th century in which scholarship became professionalized. Indeed, it has been suggested that in a discipline like history which makes extensive use of ordinary language, the use of footnotes is the only clear way of distinguishing the work of a professional from that of an amateur.

Professional specialists, being familiar with the level of scholarship and the documents in their particular fields, are the judges of achievements in these fields. Because the value of a professional researcher's work is determined only by the members of his professional group (academic association or university), the specialist charts his research in ways that he takes to be generally approved by the specialist group to which he belongs.

The old-style scientist who was active in the academy addressed himself to intellectuals in general and put some thought into pleasing as well as informing his audience. (This ethos still remains in the humanistic disciplines.) Like the public demonstrations conducted at London's Royal Institution in the early 19th century, his experiments were ingeniously devised to demonstrate his theories with flair and to sell himself and his work to society.

But once the general public accepts the validity of the scientific enterprise as self-evident, professional specialists no longer need appeal directly to society. Recognition by other specialists is sufficient. In order to acquire this recognition, however, scientists tend to pursue work that their fellows will find meaningful. Big theories and grand experiments may be impressive, but work that other scholars find useful is more highly evaluated. For instance, someone who succeeds in working out the coefficient of expansion of copper wire correctly to one more decimal place may be awarded a Ph.D for his work. This achievement may not be particularly significant even for the author's own scholarship, but since sooner or later someone in the field will almost certainly make use of it, his peers consider it worthy of recognition. In this context, attempts to force one's pet theories on others are politely ignored while "objectivity" that makes one's work intelligible to anyone is held in high esteem.

The eyes with which the professional scientist looks at nature are no longer those of the ancient natural philosopher. They have been transformed by paradigms, by specialization, and by professionalization. The result is what might be called a "professional view of nature," or rather several of them. The mechanical physicist looks at the world as matter and motion, the technologist sees it as something to be exploited, and the ecologist regards it as something to be preserved.

As these professional scientists continue to turn out massive amounts of literature day after day, the problems on which they work become increasingly subdivided and specialized. This leads to more sharply defined problems, but insofar as these problems are divorced from the larger issues of the world and human existence, and remote from the concerns of everyday life, the scientist finds himself engaged in a very exclusive enterprise.

Those engaged in this enterprise justify their work with familiar slogans: "Science for its own sake" or "Learning for learning's sake." These words were sometimes employed to assert the independence of science in the face of church intervention, but they have recently been more widely used to defend the ways of the ivory tower species against society at large. To my knowledge, these words were first uttered in 1848 by established German scientists seeking to counter the student movement; and, in the light of what has been said about the emergence of professionalism in the 19th-century German university, this is not surprising. To the contemporary specialist, however, "science for its own sake" is simply science of, by, and for the scientist.

The German notion of *Wissenschaft* is intimately bound up with the whole of German university life. It has even been said that the most appropriate definition of the term is simply "everything that is studied at the German university." But *Wissenschaft* was also the educational ideal of early 19th-century natural philosophy. One can, of course, enumerate, as I have done, the conditions which favor the admission of any particular discipline to the university and its eventual acceptance as *Wissenschaft*. Yet in the long course of history, one must assume that there has been a great deal of happenstance involved as well. This may have been particularly important in determining the shape of German *Wissenschaft*. It did not emerge fullblown from a government drawing board but, as it were, half spontaneously, under a power balance among the several pre-established disciplines.

Internal Decay

There can be no such thing as a professional in a scientific or academic revolution: a professional, by definition, follows the standards of conduct of a profession, while revolutions involve a departure from these standards.

If an engineering professor were asked what kind of invention he was working on, his initial response would probably be a wry smile. Knowing that to allow himself to be thought an amateur might damage his professional reputation, he would probably set the record straight by pointing out that he is a trained specialist who conducts research systematically—not an ordinary tinkerer who plays his hunches like a prospector, hoping to hit upon some new idea. For the same reason, the professor will confine himself, in effect, chiefly to the work of normal science, to fact-oriented research and the refinement of existing kowledge. Although this professionalization of scholarly activity has made it increasingly difficult in recent years for those outside the profession to generate scientific revolutions, a careful look at the matter indicates that revolutions commonly originate not with those who work within a single discipline or scholarly tradition, but with scientists operating on the boundaries between two or more of them.

Inasmuch as education and research are inseparably bound up with one another in the university, senior scholars ordinarily choose a successor from among those they have taught. In doing so, they usually select people who are content to work within the framework of their own paradigm. The old cliché that in dividing his territory among several disciples a scholar is engaged in quantitatively progressive, but qualitatively regressive, reproduction does not perhaps bear repeating, but given the group replenishment structures of the university, the process itself is well-nigh inescapable. Exceptions can be found, but they are unusual cases. Latter-day successors find it more difficult to break away from received traditions than their forebearers because they have not been involved either in the experience of creating a particular mode of learning or in making a place for it in the academic world. From the very beginning of their careers, they have been brought up within a set course and trained to pursue it.

Gradually these epigones create structures shaped to suit those within the university, closed structures that shut out (or tune out) external challenges. The human contacts within these walls are also limited, being confined largely to those in the same specialty. Because the external structures are perpetuated intact even though the scholarly activity going on within has lost its capacity for development and started to decline, internal rot eventually sets in. Academic logic itself cannot raze this institutional husk. Short of the radical institutional reforms that accompany political revolutions, business continues as usual. Still, if regressive reproduction results in dwarfing, there is a danger that a radical turn of events might lead to extinction, for evolutionary law indicates that overly specialized organisms lack the capacity to survive in the face of sudden changes in the environment.

Successors and Choices

When someone puts forward a truly revolutionary idea, the immediate reaction of his colleagues is likely to be rejection. At best they can be expected to be skeptical, not necessarily from ill will.

There is psychological resistance to jumping at and becoming involved in the development of a revolutionary paradigm. Beyond this, they may simply be past the age when they take an interest in new paradigms and are slow to appreciate their revolutionary character. There are times, of course, when a new paradigm is immediately given a favorable reception. But even on these occasions an established scholar whose patterns of thought have already been formed must deal with it within his own framework. Thus, even though the new paradigm presents a challenge to these patterns, the mature scholar normally has neither the desire to change his approach nor the capacity to do so even if he wished. Witness the tragedy of classical physicists trained in the 19th century who were unable to break down their hard billiard-ball image of the atom even when confronted with the early 20th-century challenge of atomic physics and quantum theory. It is, in short, members of the younger generation who first develop the work of paradigm-makers—not their predecessors or their peers.

When scholars put forward their theories they expect that somebody will listen. They may not be interested in promoting their ideas to the general public, but if scholars never took note of what their colleagues said there could be no shared scientific activity. Ordinarily, scientists think primarily about gaining the recognition of their seniors. Success at this level, of course, bodes well for career advancement, for it probably means a good job and perhaps an award of some kind as well. From the point of view of scholarly traditions, however, it is the younger generation's acceptance that is really important. Even if they could appreciate it, the older generation is hardly in a position to carry on and develop new work. Unless some members of the younger generation take it up, the fruit of the scholar's labor will die with him, he will have contributed nothing to any scholarly tradition. The older generation can, after all, only appreciate new work within an established framework, so that the work that seniors do acknowledge is apt to be of the mopping-up variety.

Mendel's law of heredity went unnoticed when it was first made public. The contribution to scholarly tradition it could have made was lost during the several decades before its significance was finally appreciated. One may speak of the problems involved in communicating scientific achievements and the "time lag" between Mendel and the academic community of his day, but part of the responsibility lies with the younger generation and its failure to accept the Mendelian challenge.[2] That a system which conforms to the natural growth and development of science is preferable to the tradition-bound establishment that was at work here should be self-evident. But it is hard to find historical examples of institutions or organizations in which the younger generation has actually sat in judgment on the work of its elders.

A comparison of the ways in which England and France sought to encourage invention during the industrial revolution suggests a related observation. In England inventors were backed by entrepreneurs who looked upon them as "golden eggs" and jumped at any chance to exploit their ideas for industrial purposes. Boulton's support of Watt is perhaps the most famous case in point. In France, a socioeconomic base conducive to the support of inventors had not yet fully matured, but the state traditionally had a policy of patronizing and developing prospective inventors. Inventors submitted their work to a screening committee composed of senior scientists, and many of the best ideas were awarded prizes. Yet the system was plagued by the bureaucratic character of the committee and the prevailing attitude of inventors who tended to think that the sole purpose of the exercise was to receive the prize money and attendant honors. As a result, France never achieved the results that the English did with their more informal approach to selection, support, and development.

Achievements of paradigmatic magnitude are commonly the work of the younger generation. In the scientific world, creative work is often concentrated in the "golden" thirties; this is particularly true of paradigm-building (Lehman, 1953). Normal science work also seems to peak in the thirties, but much can still be done in one's forties and fifties. In other words, the scientist gets his basic ideas while still young and works thereafter to explore and develop them. This means that a paradigm developed by someone in his thirties is likely to be supported chiefly by those in their twenties. But these young scholars are not mature enough

[2] The usual explanation for the fact that Mendel's work went unnoticed by scientific circles for so long is that Mendel was leading an isolated monastic life and had little contact with the outside world. Actually Mendel did make an effort to communicate with the academic community, but his findings failed to match the contemporary "sense of problem," and his work was simply ignored.

to assess the work of those a decade ahead of them, and they carry little weight in the academic world, while those seniors who really count are apt to be promoters of normal science. The number of scientific revolutions that have been aborted under such conditions is probably much larger than we imagine. Scientific revolutions that never got off the ground have obviously not been depicted in those histories of science which are nothing more than showcases of past achievements, and materials necessary for this kind of research are hard to come by. Nevertheless, the search for mechanisms which abort scientific revolutions prematurely is one of the major challenges which now confronts the history of science.

The best way to succeed in the academic world is doubtless to do a good job on some normal science under the guidance of an accomplished scholar and get oneself established while still young. Having done this, one can set off on an intellectual adventure following one's own ideas. Are there not a great many who struggled hard and long under the weight of normal science conventions only to find when they had gained status that their ideas had gone stale and they no longer had the energy for venturing further afield? Of course, unless one attains proficiency in normal science, he cannot participate in the cumulative progress of science. At what period in a career one should embark on intellectual adventure is an extremely important question both for individual scholars and for the fate of academic traditions—one that might be fruitfully pursued by studying actual cases. Such studies could, among other things, help to identify some of the external factors that intimidate the young.

Reclassifying Knowledge

Coincident with the professional specialization of the 19th century, there was a great deal of talk about classes and classifications of knowledge. This discussion was stimulated by the restructuring of university faculties and departments brought on by the invasion of the new sciences, but even more finely structured classifications were made necessary by attempts to conceptualize and define what a particular discipline was all about. Philosophically-minded writers on science such as Ampere, Comte, and Spencer were obviously interested in the problem, but even ordinary scientists found themselves drawn into the debate. It confronted them whenever they sought to relate the new discipline they were developing to what was already being taught at the university, whenever they became involved in departmental problems, and whenever they fought over the distribution of faculty positions. No one seemed to be without an opinion on the question. Moreover, as the literature brought forth by these new scholars began to accumulate, major revisions were made in library classification systems as well (Edwards, 1859).

Why is it, then, that something which caused so much excitement in the 19th century has not been much of an issue in the 20th? The first reason is that once university faculties and departments were reorganized and science securely situated in the university, the matter seemed settled, and the system that had evolved seemed fixed and permanent. Secondly, as the growing complexity of normal science widened the gap between education and research, advocates of new modes of learning began to look elsewhere for room to develop rather than investing their energies trying to crack the hard university shell. With the coming of the 20th century, scholars started to think in terms of some new kind of institution that might serve as an alternative center for scientific growth and development, and soon the university was no longer the only place where pioneering research was taking place. There were, for instance, specialized research institutes where scholarly activity went on unrestrained by systems of classification rooted in the faculty-departmental composition of traditional institutions of higher education. These institutes and their projects seem to have led scholarly minds in other directions and perhaps even relieved some of the pressures that fueled the earlier debate—though undergraduate education itself is in many respects still not free from faculty-departmental arrangements that follow 19th-century systems of classification.

The configuration of university faculties and departments that fell into place by the end of the 19th century was, then, much like the one with which we are familiar today. These faculties and departments can be regarded simply as convenient administrative units, but clearly they have other dimensions as well. One could observe, for instance, that the old faculties of the medieval university—theology, law, medicine, and philosophy—

still function as replenishment mechanisms for the traditional professions (clergy, lawyers, physicians, and teachers) and serve the needs of their professional guilds. Departments such as astronomy and physics, on the other hand, are related to the new research professions that originated in the 19th century, each with its peculiar paradigm, subject matter, methodology, and academic society.

Upon close examination, however, one sees that departmental divisions are not necessarily built upon a thoroughly consistent classification system. Looking at the range of fields in the natural sciences—mathematics, astronomy, physics, chemistry, biology, geology, etc.—one notices certain discrepancies. Mathematics lacks the type of subject matter generally associated with a natural science, while subject matter is the defining characteristic of astronomy, biology, and geology. Physics and chemistry are distinguished by their methodologies. What I have described is, in short, not a system but a sequence, a sequence which happens to follow the chronological order in which paradigms were established in these fields and they became finished disciplines. The preferential position accorded to age in this "seating arrangement" still permits us to see the historical process in which new modes of learning edged themselves into the university on top of those that were already there, and suggests that our 19th–20th-century assemblage is less a product of rational thought than of intra-university struggle and compromise among new and old academic professions.

During the first half of the 19th century Auguste Comte did attempt to develop what could be considered a rational foundation for this arrangement. His "hierarchy of sciences" parallels our historical sequence (mathematics, astronomy, physics, chemistry, biology, sociology), but it is designed to reflect certain other principles as well. It begins with the abstract and general disciplines and moves toward the concrete and particular—moving at the same time from disciplines that deal with relatively simple subject matter to those which treat of the more complex. Where the subject matter is simple, abstraction and generalization were quickly achieved and paradigms established early. Comte's hierarchical order agrees with the historical order in which these scholarly disciplines were established and bears at least superficial resemblance to the make-up of modern university faculties and departments. His scheme has the same problem we met at the at the outset of this section: the conflict between the older Aristotelian modes of learning which followed a division of knowedge based on subject matter (physics dealt with natures below the moon, astronomy with those of the moon and beyond), and the modern sciences distinguished by their methods. The older studies, then, are not really parallel to the newer methodologically-grounded disciplines.

The situation seems to call for a structural division wherein a field that deals with the heavenly bodies using the methods of physics is known as astrophysics and one that deals chemically with living things is called biochemistry; for departments set up on the older subject-matter principle seldom seem to generate original methods or paradigms. However the world may have looked in Comte's time, from the vantage point of the 20th century the history of modern science could even be written in terms of the invasion of subject-matter fields by the methods of physics. It is this encounter of dynamic methodologies with departments rigidly defined in 19th-century terms that has created tensions and contradictions in the 20th-century university.

In the teaching of preschool and elementary school children, subject matter classifications are still meaningful and may always be so, but a switch to a methodologically ordered curriculum should be made at the middle and high school levels so that students can gain experience in handling scholarly paradigms. Of course the quest for knowledge does not end with the learning of paradigms. Even though received paradigms should prove capable of resolving most of the problems of normal science, unresolved subjects for investigation still remain to challenge us (e.g. earthquake prediction, the nature of life, etc.). It is to be hoped that these problems will be apportioned to research institutions and that scholars with expertise in the relevant plural paradigms will be engaged to work together on them.

Industrial Society and the Place of the Technician

The practical arts were not part of the orthodox academic traditions of East or West. In both worlds those who could handle literary culture were regarded as superior to those who made things with their hands—possibly because letters long were used as a primary instrument of bureaucratic control. In Europe as elsewhere, artisans were normally illiterate. An artisan's line from a comic dialogue of the Edo period reflects an attitude that was no doubt more widespread: "That fellow's hankerin' after book learnin'; he must be a clumsy hand."

If this identification with instrumental techniques was part of the artisan's self-understanding, it also set the terms of his relationship with his social betters. In the eyes of the government official the artisan was less a user of techniques than someone to be used for them. Thus the author of the *T'ien Kung Kai Wu* of the Ming period cautions readers in his preface that the information he has collected has nothing to do with improving one's station in life. Gregory King has estimated that among the thirteen classes distinguishable in the English population of 1688, skilled artisans ranked fourth from the bottom—above soldiers and sailors, seamen, agricultural and industrial laborers, and impoverished farmers, but below tenant farmers and shopkeepers—in the class order.

Yet there were technicians who enjoyed a somewhat higher status. These were experts in military technology, men like Leonardo da Vinci, who began to appear in significant numbers during the Renaissance. Well versed in military tactics, these strategists were skilled in building fortresses, encircling them with moats, and making use of guns and cannon. Finding favor with princes, they were well paid and well treated, and in peacetime they sometimes did the type of work we now associate with civil engineers. When armies became complex organizations in the modern period, these functions were absorbed by an "engineering corps." Thus the word "engineer" originally referred to this kind of military technician.

The first country to create a modern educational system for the training of military technicians was France. Colbert and Vauban had undertaken to modernize military technology and develop a military corps of engineers under Louis XIV, but it was in the 18th century that the training of military technicians became fully institutionalized. After a decade of major reforms (1765–75) the École de Mezieres entered its most prosperous years, and, during the Revolution, the organization and talent it had created were inherited by the newly established École Polytechnique. Military and civil engineers were trained in government schools and worked in government institutions with government support so that they were the equivalent of middle-level government officials. Meanwhile, private-sector technology was kept alive by artisans, trained as of old in the apprentice system and situated as before near the bottom of society.

Technology and Higher Learning in England and France: The Development of the Steam Engine

When one follows the development of the steam engine from its first beginnings in the 19th century through the contributions of Savery, Newcomen, Smeaton, Watt, Hornblower, Murdoch, Woolf, Trevithick, Southern, and finally Stephenson, one discovers that almost every major step forward was the work of an Englishman. The government of England, however, had nothing to do with its development. Toward technology it maintained the same *laissez-faire* policy ("we have nothing to do with it") that marked its attitude toward university education and the scholarly activities of the Royal Society. In the age of mercantilism, scientific navigational techniques (e.g. longitudinal measurement, development of the chronometer, etc.) had been considered matters of national interest and received some government support, but the technology of the Industrial Revolution was technology for private profit and personal success. In the latter half of the 18th century patent letters issued in promotion of technological growth replaced those granted as royal favors or concession; but beyond this, technicians of the time did not find governmental intervention or assistance necessary.

What kind of education did these English technicians have, and what were their attitudes toward formal learning? First, none of them had received any higher education. Coming from Nonconformist backgrounds, they were outsiders by birth, kept at a distance by the upper classes who went to Oxford and Cambridge. Even when they succeeded in accumulating some money, they were not in a position to buy land, manage estates, and become part of the high-church Anglican upper class. As a result, they cast their lots with that vulgar young upstart—technology—and with factory management.

Going to a university was not a possibility, but a university education would not have been of much use to them in any case. Technological skills were learned exclusively in on-the-job apprenticeships. Smeaton and Watt did place announcements in the *Philosophical Transactions of the Royal Society,* but they were exceptions. The average technician had no contact with scholars and no desire to seek notoriety by reporting his inventions in a learned society's journal. Since they were solely concerned with obtaining patents and putting their inventions to enterpreneurial use, the only materials the technicians left for the historian of technology were patent applications and the products they manufactured. Samuel Smiles, who became so familiar to Meiji Japanese youth through Nakamura Masanao's translation of his book *Self Help,* wrote a number of biographies of this kind of technician, and these are still of interest to the historian of technology today. All of them depict the realized Victorian dream of success through self-reliance.

These technicians probably did make use of manuals and handbooks like those written by J. Southern and Oliver Evans, but these were hodgepodges of experiential knowledge: information about the price of a machine that could lift water out of a mine so many feet, or how to fix a leaky cylinder by plugging it with felt from an old hat, and so on—quite remote in style from academic writing.

In the French industrial world, the demand for the steam engine was still largely latent. The first steam engine in France is said to have been imported from England and used to pump water in the court fountain. Under a policy of patronizing technological development that stood in marked contrast to English attitudes, men worked for prize money, competed for fame, and sometimes set their sights on a noble title even when (as was often the case) their inventions were not practical. The original ideas of Papin and Cugnot notwithstanding, the French contribution to science lay not in the institution of government prizes but in the much more basic area of developing a concept of engineering that effectively met the needs of the times. Embodied in the École Polytechnique, this concept was to be the fountainhead of modern engineering science.

Scientists and technicians trained in the Polytechnic tradition were more interested in the mechanism of the steam engine than they were in its potential uses. Several papers concerned with the steam engine were read by leading scientists and technicians at the Académie des Sciences, including the extremely laborious theoretical presentation of De Pamaour, in which such things as cylinder diameter, volume, and piston velocity were submitted to aerodynamic analysis. It was in this ethos that the pioneering work in thermodynamics of the Polytechnician Sadi Carnot would eventually emerge.

Science and Technology

The relationship between science and technology is another theme that once attracted the attention of Marxist historians of science. Their interest in J. D. Bernal's *Science and Industry in the Nineteenth Century* is one manifestation of this concern. Bernal set out to demonstrate that what we call science was not immaculately conceived within the ivory tower, but rather was born from, and in conjunction with, technology. In undertaking this task, he was seeking to challenge the common notion—particularly pronounced in traditional English intellectual circles—that whereas science was the work of upper-class gentlemen and had emerged from within the establishment, technology was something much more base.

In reality, the two traditions were fairly distinct prior to the 19th century. If there was an exception, it was in France, where the government had long encouraged the Académie des Sciences to conduct projects aimed at the practical application of science. One cannot say that these attempts to fuse science and technology always went well, but they did contribute to an atmosphere in which the two were closer than in England (Hahn, 1971). Pre-Revolutionary France also had many more institutions designed chiefly for the training of military technicians than any other country—a school of military engineers, an artillery school, and a school of bridge and road construction, all of which offered a formal scientific education. It was as an extension of this tradition that the École Polytechnique and its conscious effort to combine science and technology emerged during the Revolution. Science was now to be the basis for technology, and technology the application of science. This new relationship, however, did not proceed in an uninterrupted straight line.

The Establishment of the École Polytechnique

In the days of "great plans and meager results" that marked the first years of the French Revolution, wide-ranging debate on higher educational policy failed to produce a working consensus. There was agreement on the destruction of the old system and the creation of a new one, but with respect to the character of the new, the position of science and technology, and the relationship between lower and higher education, there were a variety of conflicting opinions. Some, like Nicolas Herz, argued that higher education should be limited to training in technology and military affairs, while Georges Bouquier and others called for public recognition of the fact that applied science was grounded in pure science. Behind this controversy one can distinguish two groups, "technocrats" from conventional military and technical schools, and practitioners of the new sciences and proponents of the Enlightenment. The Jacobin position on this matter is often said to have been one of unbounded support for technology and aversion to science, which they saw as dominated by aristocrats (Gillispie, 1960). Yet the Jacobin critique of science did not necessarily involve a clear distinction between science and technology. The general view seems to have been that once science was institutionally removed from aristocratic control, the rest could be left to the scientists (Fransson, 1968; Nakayama, 1974).

Inaugurated in 1794 as the École Centrale des Travaux Publiques (the name was changed the following year), the École Polytechnique owed a great deal to the vision of Gaspard Monge. J. Lamblardie was installed as its head, but the content of the curriculum and selection of instructors generally followed Monge's ideas. Several outstanding scientists from the old military-technological schools were included in the faculty, but the École Polytechnique was to be different from these schools in two ways:

1) the study of applied sciences such as civil engineering and fortifications was to begin after students had acquired a solid foundation in mathematics, physics, and chemistry; and 2) attempts were to be made to closely link basic theory and application. The first issue of the school journal, the *Journal de l'École Polytechnique*, contained detailed accounts of the ground to be covered in each course, while a *Tableau Analytique* charted each subject and its place in the curriculum. There is clearly a conscious attempt here to put together a systematically structured body of studies. In the military-technical schools, even the textbooks had been more loosely structured. For example, *The New Principles of Gunnery* (1742) of artillery school professor M. Benjamin Robins begins in the very first chapter with a discussion of artillery ammunition, touching upon basic science only from time to time as it goes along. The École Polytechnique was the first place where the basic sciences were the foundation for a course of instruction in technology.

When the École was founded, courses in basic science took up one-third of the three-year course, with applied technology making up the other two-thirds; but soon the number of basic science courses began to increase.[3] Behind this development lay the growing power of the military and the revitalization of the old military-technical schools, which now began to take over applied education and transform the École Polytechnique into a preparatory school for their own institutions—opening a gulf between science and technology once again. With the advent of Napoleon in 1799, this process was completed. Students wore military uniforms. The Republican vision of the school as a training ground for industrialists, teachers, and scientists was eclipsed. "L'École Polytechnique is dead," Monge is said to have lamented in his last years. But with its "death" a new type of education was born—a strict, military-style, intensive course of instruction in the natural and applied sciences that contrasted sharply with the university tradition of rhetorical learning.

Technology as Higher Learning

What does technology look like when it comes to be taught in

[3] Fourcy (1928) contains a table of the curriculum and various statistics.

schools? Peter Drucker (1961) has outlined the process which led to modern-style academic technology in four stages: 1) the collection of existing knowledge; 2) the organization of this knowledge; 3) its systematic application; and 4) publications. This scheme, however, only accounts for the emergence of Baconesque collections of practical knowledge or repositories of technical information like the French Encyclopedia. In the previous chapter, I argued that modern science was the product of a successful transformation of practical traditions of knowledge. I would now like to suggest that what we call modern "engineering" is a body of scientifically reformulated and reoriented techniques that resulted from the reintroduction of this modern science into the world of practice. Just as the rise of modern science marked a parting of ways between East and West, so the emergence of modern engineering created a decisive technological gap between the two cultural spheres.

The modern engineer can be distinguished from the craftsman by virtue of his sensitivity to science and the scientific character of his work. Traditional technology was learned by experience and transmitted half-secretly in craftsmen's guilds. To have made it public would only have been to invite theft by other groups of craftsmen. Even in England, where the industrial revolution was most advanced, patent laws designed to protect techniques did not exist before the last half of the 18th century, and technicians remained reluctant to publicize their inventions (Booker, 1963). A similar atmosphere prevailed in France, where Gaspard Monge set out to gather materials from artisans and military technicians and publish his book on descriptive geometry—only to have publication banned and the work classified as military information.

But the ban was not to last forever. Modern science had grown up in an atmosphere which considered publication of one's work a universal obligation. Moreover, open access to knowledge was basic to school education. Against this background the demand for technical education eventually led to the publication of Monge's *Géométrie Descriptive* in 1795. France was at the time under Thermidor control; Écoles Centrales had been established throughout the country, and revolutionary educational reforms had been initiated. These reforms brought new educational opportunities to the sons of artisans and craftsmen. The demand of these young men for courses that would be of immediate use made drafting more popular than mathematics, the natural sciences, or any other subject.

If Monge's work met this demand, it also presented the drafting techniques of humble artisans in a manner sufficiently elegant to attract the attention of intellectuals. Although his book did not provide a great deal of new information, it did lay it out more clearly and accurately than before.

Descriptive geometry stood at the historical divide between craft techniques and modern engineering. Even today, the popular image of the engineer as a man who walks around with a T-square or a slide rule remains alive. Indeed, inasmuch as drawing continues to be regarded as the lowest common denominator of a modern engineering education, the use of sophisticated drafting techniques may still serve to distinguish engineering technology from traditional craft techniques.

To single out drafting as the special feature of modern engineering may sound overly disparaging. Yet it was drafting that effected the translation of techniques into mathematical terms (standardization and quantification) and gave modern engineering a new orientation. And this was the beginning of its enlistment in the mass-production factory system of modern industrial technology. In Monge's day, descriptive geometry had a broader meaning than it does today.

The descriptive geometry that along with mathematics, physics, and chemistry composed the technological curriculum of the École Normale (where Monge had taught earlier in his career) included application to such things as traditional stonecutting and construction work, machine principles, and many things that are now considered part of mechanics.

Writing in a somewhat broader context, Hirosige Tetu (1973) aptly captured what is distinctive about the method of modern engineering when he described it as a discipline, or rather a series of disciplines, built upon unit-based systems and a factor approach to analysis. It therefore cannot be adequately understood in terms of the *Wissenschaft* and academic science of the German university. Rather it should be perceived against the background of the French École Polytechnique (the heart of "industrialized science") and the radical quest for scientific rationality that was a part of the

French Revolution—a rationality that also produced the metric system, the decimal system, and the 24-hour clock.

Thus it was with the establishment of the École Polytechnique that modern engineering was really born. Working with Monge, men such as Coulomb, Lazare Carnot, and others produced a generation of successors like Poncelet and Coriolis who took the lead in the construction of modern engineering science. Over and against the abstract analytical mathematics of Lagrange, Laplace, and Cauchy, these men formed a school of applied mathematics that was designed for application to the problems of weaponry, mechanics, and architecture. Linking Monge's projective geometry with the mechanics of machine technology, they worked with physicists and chemists to develop a modern science of engineering. Poncelet's lectures served as a model for the textbook used in the German Technische Hochschule and eventually influenced engineering education throughout the world.

Academic inquiry in the German university proceeded by a variety of methods, but in the École Polytechnique the machine was the model. Things were broken down into their elemental parts, measured, reassembled, and then adapted. Through the thorough application of this approach, the intuitions of the traditional craftsman were converted into science. The view of science implicit in the notion that it is more scientific to measure the temperature of the bath water with a thermometer than with one's hand began to spread—even into the everyday activities of family life. At the same time, materials that were readily measured and factual knowledge about them came to be regarded as more scientific than, and hence superior to, other fields of academic inquiry.

The organization of an early engineering textbook, *A Manual of Machinery and Millwork*, published in 1869 by one of the pioneers of modern engineering, William Rankine, underlines the centrality of drafting in the new disciplines. Its chapter headings—Mechanical Geometry (drawing); Machine Dynamics (Newtonian mechanics); Materials, Strengths and Structures—also suggest that, in addition to the general textbook progression from easy to difficult and simple to complex, modern engineering was, at the time of its birth, structured in terms of the paradigmatic principle of movement from the basic to the applied. Its foundations lay in Newtonian mechanics and in physics and chemistry. Only drafting—that is, the geometry of machinery—might be considered original.

But what happened to technical skills when they were transformed into the modern academic discipline of engineering? First, they were emancipated from secretiveness and raw experience, and took on a universal character. This universality was very important, for it made possible a standardization of techniques that led, for instance, to the manufacture of interchangeable parts. Second, in providing technical skills with a theoretical base, the academic transfiguration of technology gave modern engineering and modern technicians a social status higher than that of the artisan. Still, engineering was not the product of an advocate group committed to a unique paradigm. If there were paradigms involved, they were those of modern science. Engineering came into being as a joining of the paradigms of modern science to the collected pre-paradigmatic skills and experiential knowledge of artisans. In the language of this book, it does not seek to generate paradigm change, though it can anticipate normal-science development.

On the other hand, technology can be seen to have paradigms and revolutions all its own. To take the case of electricity, during the 19th century Edison's direct current competed with Westinghouse's alternating current until the latter succeeded in acquiring a more substantial group of supporters (in the form of financial backing) and became the major means of generating and transmitting electricity that it still is today.

The Technische Hochschule and Modern Engineering

Born in the École Polytechnique, modern engineering education grew to maturity in German-speaking lands. Under the Napoleonic regime, the once novel French educational system turned rigid and resistant to change, but in Germany the territorial princes contended with one another to see who could build the newest Technische Hochschule (institute of technology) with the most up-to-date aims and curriculum. Engineering education flourished in countries whose industrial revolutions began only in the mid-19th century, partly because scientific technology matched the needs of a society in the throes of industrialization, and partly because of the same inter-state rivalry that promoted the growth of the German university.

Would it not have been better to create an engineering faculty within the university than to build separate institutes of technology? This problem was discussed in the 19th century and has been an issue ever since. That the Technische Hochschule was kept apart from the university is perhaps best understood in terms of a subtle but unmistakable defensiveness that began to creep over the German university as the Revolutions of 1848–49 passed into history and the last half of the 19th century arrived. Dominant opinion feared that the university's commitment to "learning for learning's sake" would be compromised by the introduction of such utilitarian subjects as engineering and agriculture, and institutional growth in that direction came to a halt. In 1899 a movement of technicians allied with the industrial bourgeoisie did result in the Technische Hochschule's being given a status equal to that of a university (including the right to grant academic degrees), and today some of them call themselves Technische Universität, but their historical isolation from the basic research of university philosophy faculties (and the independent natural-science faculties that have grown out of them) is acknowledged. Intellectuals have regularly criticized this isolation, but it is also possible to argue that it was precisely the historical separation of the Technische Hochschule from the tradition-bound university that enabled modern engineering to acquire a systematic shape and permitted it to develop. In any event, the Technische Hochschule, like the German university, became a model for engineering education in several other countries.

In the early 19th century a group of English scientists established a Mechanics Institute to provide scientific education for lower-class artisans, but the venture proved unsuccessful, largely because of a scarcity of teachers. Modern engineering education was not to take root until a system of compulsory education had been founded and staffed by teachers who had received university-level scientific training. As we have seen, however, making provisions for this training was in itself no simple matter. Even at mid-century the old English universities refused to open their doors to engineering, though many observers at the first World's Fair in 1850 had come away from the exhibits with a strong feeling that

British technological superiority was threatened. As a result, institutions that concentrated chiefly on practical knowledge were founded or expanded in London (London University) and several other regions.

Confronted with the need to raise the general level of technology, the English governing class's first response was the construction of technical schools that would, by equipping the poorer classes with technical skills, kill two birds with one stone—promoting industrialization and alleviating poverty at the same time. This policy reinforced a habit of thought, particularly noticeable in England, which looked upon science as an upper- and middle-class intellectual activity and technology as a lower-class occupation. Consequently the social standing of engineering changed but slowly despite its achievements.

In places that were relatively free from tradition-based prejudice, engineering education extended itself more easily. In Scotland, technology found its way into the university at an early date. Russia established higher technological institutes in St. Petersburg and Moscow under the influence of German culture. The St. Petersburg Institute of Civil Engineers even included a "training factory" where students could acquire practical experience, an educational formula quickly adopted in the United States and in many other countries.

Inaugurated in 1865, the Massachusetts Institute of Technology's "learning by doing" system added heavy doses of laboratory and homework to the foundations laid by the Technische Hochschule. In contrast to the German system where the student was at liberty to attend whichever lectures he chose, MIT students were given intensive training in a fixed basic curriculum. The type of freedom accorded the 19th-century German university student had demonstrated its value in the humanistic disciplines, but it was not an effective way to bring large numbers of students to the frontiers of normal science in technological traditions that were constantly on the move. For this reason, an ultramodern carrot-and-stick approach to education began to spread across American higher education.

In the meantime, Japan had made its appearance in the 19th-century world. The Japanese government had asked Henry Dyer, a twenty-four-year-old disciple of Rankine, to develop a facility for the teaching of engineering. Dyer set out with eight associates to construct a comprehensive engineering curriculum that combined science and technology. Taking as his model the Swiss Institute of Technology, opened in Zurich in 1855 and then the newest such institution in Europe, he sought to link his native English practicality with a theoretical dimension still unknown in England.

When the Imperial University was established in 1886, it absorbed the fledgling school Dyer had helped to create, and engineering took its place in the university alongside law, medicine, and the liberal arts and sciences. Japan was perhaps the first country in the world where this occurred. (With the exception of Scotland, it has still not happened in Europe.) Moreover, according to Amano Ikuo (1969), in 1890 the Imperial University's engineering faculty had a higher percentage of *samurai* (upper-class) graduates than any other (engineering 85.8%, natural sciences 80.0%, law 68.3%, medicine 40.8%). These figures suggest that Japan's leadership class as a whole (not just government policymakers) was relatively free of the kind of prejudice against engineering and technology that characterized 19th-century England. In other words, as non-Western Japan looked out on the world through freshly opened windows, it appeared uninterested in the traditional status differential between science and technology: to the men of Meiji, both seemed equally tools of civilization and enlightenment. The year 1890 did see strong academic opposition to the addition of an agricultural faculty (including departments of agriculture, forestry, and veterinary medicine) to the Imperial University, but resistance finally bowed to bureaucratic determination. In insisting on the importance of agriculture, the government was acting in a manner suggestive of Japan's position as a newly emerging nation, for new states characteristically emphasize applied fields that have an immediate bearing on local problems.

The Origins of the Technocrat

As the starting point of modern scientific and technological education, the École Polytechnique has won many accolades. It was here that modern scientists were recruited as professors for the first time. There had been scientists who were professors previously, but they had usually entered science from within the university system. Only with the École Polytechnique do we find an institution which was conceived as an agency of modern scientific and technical education.

The École Polytechnique has, however, not been without its detractors. If young people spend the first two years of their higher education having their heads crammed full of nothing but mathematics, physics, and chemistry, and their third year learning how to apply this knowledge, what kind of human beings will they become? In the case of the École Polytechnique, F. A. Hayek (1952) has argued that it turned out men with a propensity for dealing with society by the methods of natural science and engineering, men who thought human and social reconstruction was as simple as building a bridge or a road. In their pursuit of quantified reason, they tended to have little appreciation for things like human alienation and liberation that resisted quantification. It was these *polytechniciens* who filled the ranks of the Saint-Simonians, a group that Engels referred to as "utopian socialists."

Of course this faith in the power of science was not peculiar to Polytechnique graduates. The curricular reforms of the French Revolution ignored or discontinued classical languages, treated literature, history, and grammar as inferior subjects, and abolished all religious instruction, so that there was really little but science left (Leon, 1961). Nor was the École Polytechnique the only child of the revolutionary curriculum. If it has sometimes borne the brunt of attacks on revolutionary education, that is because it has seemed to be symbolic. But the rationalism and scientism it embodied had their beginnings under the *ancien régime* and were to continue even after revolutionary passion was gone. Under the centralized regime that succeeded the Revolution, the École Polytechnique was to produce many able technocrats. Much later the educational curriculum of the French Revolution was to become a model for socialist revolutionaries, and today both the Soviet Union and China have put together curricula that give priority to science. The desire to separate education from research (the École Polytechnique was an educational institution, while research was not yet institutionalized) in the interest of efficiency is also part of the rationalism of modern revolutionary governments.

One of the charges that is often made against this kind of system is that the technocrats it produces are never more than faithful servants of the system. They make real contributions to the development of normal science, but show little capacity for generating the revolutionary ideas and paradigms that typically originate at points quite unrelated to (or outside) the system. Between the type of thinking that is done within the system and that which emerges outside it (even in the natural sciences where the aim is "objectivity") one can certainly recognize a difference in scholarly style. Nazi science may be cited as a case in point. Under national socialism some technicians entertained the illusion that by linking themselves directly with the state and party they would avoid dependence on capitalist support and be given full opportunity to make use of their skills. They were not, as in the technocrat's dream, to become rulers themselves. But so long as they submitted to party authority and remained loyal to those in command, they found a welcome for plans that made the most of scientific technology. Thus scientists and technicians like Fritz Todt, Albert Speer, Wilhelm Kepler, and Albert Pietzsch were picked up by Hitler and put to work for the regime. During the same period, many products of the Technische Hochschule also advanced to positions of leadership.

Technology fulfills its end in being used by the system. It is compatible with policies designed to enhance national wealth and power or planned social construction, and shows little resistance to participation in controlled or planned economies, regardless of their ideological coloration. The academic science of the "good old days" had been a cosmopolitan, self-regulating world of its own. Nurtured in the bosom of bourgeois liberalism, it had served its own ends. Then World War I demonstrated the utility of science in national defense and the exploitation of natural resources, and governments began to become deeply involved in scientific affairs. Nazi Germany shared this attitude that science should serve the state, and when confronted by it many of the world's top scientists in fields with a universalistic bent and a strong spirit of international cooperation sensed a grave threat. Some of them, even non-Jews, were accused of practicing "Jewish science" and forced to seek asylum abroad. For a group of power-oriented, second-rate scientists and for those particularly concerned with local science, however, this exodus seems to have made space on the bandwagon.

When a regime is intolerant, it rejects without consideration paradigms that emerge outside the system. Thus the only paradigms born under these conditions were those produced or approved by the regime—like the ethnological idea of racial superiority. This relationship between the sociopolitical order and the generation of paradigms (as well as the growth of advocate groups) poses many fascinating problems. The nature of Soviet science, nominally purified by its liberation from capitalistic oppression, and the appearance of new academic styles in China's "Cultural Revolution" are attractive themes for subsequent analysis in this regard.

"Theatre Schools" and "Driving Schools"

Traditionally, the university was a place where instruction was given from the lecture platform. Available means of communication made this inevitable in the medieval period, but as the English philosopher William Whewell pointed out, there are other reasons for the persistence of the lecture method in modern times. It is well suited to fields in the pre-paradigm stage or in the process of paradigm formation. Indeed, for communicating the latest or for transmitting information that has not yet found its way into textbooks, there is perhaps nothing better. This "theatrical form" of knowledge transmission provided the setting for the rise of the German university's academic philosophy and the free-ranging discussions of natural philosophy that followed its lead. Many of the problems dealt with could have been better handled by the natural sciences, but these were not yet fully a part of the system. Though no longer taught through lectures on classical texts, even medicine had its anatomical theatres where professors performed with the scalpel on a raised platform, and the experimental science lectures developed by London's Royal Institution were above all "shows" for the good ladies and gentlemen of the community.

In the mechanical, as opposed to the liberal, arts—the tradition to which the École Polytechnique succeeded—training had long been given on the job, through the apprentice system. Universities might have been able to treat students simply as an audience when transmitting "insubstantial" learning, but could not afford to be so casual in inculcating practical knowledge. Should engineering be taught in a superficial way, houses would lean and bridges collapse. Thus when these manual traditions became modern engineering and entered the world of higher education, the need was seen for the development of a solidly structured curriculum in which students would be required to complete a specified course of study that began at an introductory level and proceeded gradually to more difficult studies. Tōyama Hiraku (1971) has referred to this kind of higher education as the "driving-school type." Even without using this term, the point should be clear. Unless merchant marine graduates learn enough to pilot a ship and operate a wireless, ships will sink and human lives and cargo will be lost. Thus there are educational institutions which exist solely because of the need to inculcate these kinds of received skills and techniques. It was in the latter half of the 19th century that these two distinct and historically separate strains—the university and the apprentice traditions, insubstantial and practical knowledge, "theatrical" and "driver-training"-type education—began to draw closer and closer together.

If this raised the question of how the university was to accommodate these newly elevated modes of learning and the educational methods they seemed to require, it also added a new dimension to the discussion of what to do with a mode of learning once the paradigm in which it was grounded had completed its life cycle. When a mode of learning is no longer able to generate new problems, what is its role in higher education? The knowledge it has developed will remain in the large volume of scientific literature it has produced and, after consolidation, find a semi-permanent home in textbooks, handbooks, and dictionaries. Institutionally, it will continue to be a part of the tradition until the waves of reform and revolution finally sweep it away. Yet some studies become a fixed part of the basic curriculum even though they have little significance for scholarly research and development—things like Euclidian geometry; basic mathematics for science and technology such as the differential and analytical calculus; Latin and modern languages. These are courses that must be taught because they lay the groundwork for study and research in other areas. As the educational ladder grows higher, there is a tendency for these foundation courses to be pushed down to the middle- and elementary-school levels; but wherever located, it is important that they be taught with the efficiency of a driving-school course. Meanwhile, as normal science grows more esoteric and scholars begin to speak of the evils of overspecialization, a reaction sets in. Under the name of general education, there is a revival of pre-paradigmatic, "theatrical" courses dealing with problems (e.g. theories of civilization) that are difficult to present in normal-science fashion, as either specialized or foundation courses.

The contemporary university presents many faces. It is a strange and complex mixture of specialized courses for the advancement of normal science, foundation courses for the acquisition of existing technology, and general education for the free citizen. Filled with all kinds of conflicting claims, the 20th-century university has not always lived up to the enormous expectations raised by the glory of its 19th-century German predecessors; but in the afterglow of the past, it still casts a long shadow.

The Limitations of the German University

In the latter half of the 19th century, the German university grew larger and larger and its reputation became firmly established. England, the United States, proud France, and emergent Japan all sent large numbers of students to be educated within its gates and sought to emulate it at home. In institutional terms, however, the innovative life of the German university was already over.

Strangely enough, its most creative years had more or less coincided with a period of unrest and student uprisings. Whether there was in this case any causal relationship between innovation in dispassionate, value-free academic research and a tumultuous student movement is not clear, but it seems likely that the constant criticism of the old order made for freedom of speech and a weakening of constraints on the generation of new ideas and modes of learning. In any event, things came to a head with the outbreak of revolution in 1848. The generally younger irregular faculty members and *Privatdozenten* aligned themselves with the students against the faculty councils and senates of the regular professors and the territorial governments which stood behind them. In the end, the autonomy of the university as a self-governing body was used to reject reform and preserve the privileges of the regular faculty (Shimada, 1947). Thereafter, from the latter half of the 19th century until the Nazi university retrenchment policies of the 1930s, German universities underwent almost no institutional change. Numbers of students increased and the universities seemed to prosper, but there was more conservatism and greater dependence on the state. Indeed, the institution became so inflexible that it refused to countenance the admission of new disciplines such as engineering.

Inasmuch as scientific activity seeks to create, it inevitably comes into conflict with the general orientation of institutions toward stability. By the same token, as long as science refuses to abandon creativity, it cannot be fully at home within the confines of any institution. When it does settle down, as it did in Germany during the latter half of the 19th century, shades of decline begin to appear—in the German case right in the midst of industrial rise. In general, when institutions congeal following a period of prosperity and growth, the new generation comes to be selected by the previous one rather than emerging through individual paradigm choice and advocate group formation. As a result, the reproductive process becomes predominantly a regressive one within the lineage of normal science as laid out by the senior generation. In time, the stature of newly recruited talent shrinks, and the tradition begins to show signs of aging. The German university was particularly successful in promoting specialization, and this, together with a strong commitment to research, led to great advances in normal science. The scale of research grew and the number of persons involved increased. But the university proved too inflexible to handle specialization and large-scale expansion, and they became significant factors in its decline.

At the beginning of the 19th century, levels of learning were such that Wilhelm Humbolt's idea of the unity of education and research in the university still had considerable currency. Education even served as a stimulus to the scholar's inquiry. By the latter part of the century, however, the gap between the levels of advanced research and the capacities of students just out of secondary school was so large that teaching obligations became a heavy burden on instructors desirous of pursuing serious research. Students, for their part, entered institutions where the most advanced highly specialized research was being done, only to lose their way for want of appropriate guidance—and soon to lose their initial enthusiasm for research as well. As Paulsen (1906) has observed, English and American universities seem to have been better prepared to cope at this point, for they provided an intermediate level of scientific research.

But students continued to flock to the German universities—students from abroad attracted by its reputation, and German students who, once they had passed the *Abitur* exam, could attend any university they wished. The size of the teaching staffs, being determined chiefly by the size of government budgets, could not keep pace with the increase in the number of students. Under these circumstances the German university could no longer provide the liberal setting for elite education that had formerly been possible. Every successive government heard demands for holding down the number of students to an acceptable level by imposing some restrictions, but nothing was ever done.

The German university was well adapted to the early stages of specialization, for it looked upon science largely as a matter of individual research. As scholarship grew more and more specialized, however, the traditional universities were no longer able to serve as undifferentiated centers of basic education and advanced research. It was for this reason that the baton of world leadership in university-level research passed to the American graduate school.

In the name of the freedom to learn (*Lernfreiheit*), universities had no restrictions on the number of students. With the freedom to teach and do research as they pleased, *Privatdozenten* opened up new fields of study almost at will. At Berlin University, for instance, new lectureships and lectures seemed to have appeared rather freely and disappeared with equal ease. This system worked well in a period when science was being professionalized and broken down into smaller divisions. The absence of regulations governing the size of the student body also contributed to the expansion of science. Yet as textbook paradigms of learning were established and the normal science that was built upon them advanced, the amount of basic training required before the student could arrive at the front line of normal science research increased. The researcher required the same kind of training in received knowledge and techniques that was necessary in law, medicine, or technology. In order to make this training as efficacious as possible, textbooks had to be constantly revised and brought up to date. And in order to bring students to the frontiers of specialized research in any particular field, systematic training became imperative—even if it had to be accomplished at the expense of the freedom to learn.

This was of particular significance in the experimental sciences, for laboratory instruction, unlike lecture courses, could not be given to students *en masse*. Moreover, instruction had to be given in an orderly, progressive fashion. Thus, with the introduction of laboratory courses, *Privatdozenten* in scientific fields found it increasingly difficult to make a living. With the reforms of Carl Becker in 1923, *Privatdozenten* finally became salaried employees, although this benefit was balanced by increased government intervention in university affairs.[4] But the laboratory sciences remained difficult to teach in a system governed by the freedom to teach and freedom to learn and centering around a lecture system inherited from the Middle Ages. The age of the systematically ordered curriculum, small classes, fairly intensive compulsory instruction, and the fiscal intervention by government authorities that the cost of facilities invites, had arrived. Classical liberty simply was unable to foster the development of experimental science beyond a certain point. Herein lies the origin of the transformation of scientific research into the establishment enterprise that it is today.

There are basically two answers to the type of weakness evinced by the German university: the creation of graduate schools to improve the university's capacity to educate and develop trained scholars, and the establishment of research institutes where scholars are free of teaching obligations. Toward the end of the 19th century, the United States took the graduate school approach, and a few decades later, German fears of falling behind American science prompted the creation of the Kaiser Wilhelm Society for

[4] The *Privatdozent* system had long had its public opponents, for unlike professors on the government payroll, *Privatdozenten* could not be removed even when they advocated anti-establishment ideas.

the promotion of pure research. The postwar Japanese university followed the graduate-school model, and the national research institutes of the Soviet Union illustrate the policy of separating advanced research from the classroom (Graham, 1974).

The Establishment of the American Graduate School

Until the U.S. Civil War, Harvard and the other old private American universities were places where one went to get a Cambridge-style gentleman's education. They were not engaged in opening up new frontiers of academic research. Intellectually, they had still not completely emerged from colonial status. Beginning around 1830, a group of men who had studied in Germany developed plans to introduce the German system, but these did not materialize until after the Civil War. Even then, adopting the German emphasis on research by reforming existing university faculties proved extremely difficult. At Harvard, for instance, faculty opponents argued that the creation of a research-centered system would result in a decline in the quality of education. The eventual solution was to leave the departments as they were and incorporate the German system in a graduate school. Even then, resistance remained strong in the humanities, where scholars continued to maintain that the traditional program of undergraduate education was all that was necessary or desirable. Thus, though the humanities eventually fell into line, the outfitting of the graduate school began with the needs of the new science disciplines. Consequently it conformed to new developments in the sciences in many respects. According to Hofstader and Hardy (1952), these early years of the American graduate school belong to an "age of awakening to science" (1860–1910), an age which they suggest was closely paralleled by an expansion toward specialized education that took place between 1870 and 1920. By the latter date the research-centered graduate school inspired by the German university had largely acquired the form it has today.

Almost from its inception, the American graduate school exceeded German spending on research facilities and equipment, and some thought was given to financial aid for graduate students. Moreover, unlike European-type postgraduate training where individual study commenced immediately, American graduate schools put students through rigorous training in a basic curriculum during the first half of their residence (approximately two years).

Being both one step removed from the world of commercial incentives and unwilling to leave things wholly to student initiative, American graduate schools used artificial incentives and a merit system to raise the level of scholarship. "Carrots"—a scholarship system where funds were awarded in relation to one's marks and salary increases given to those who had gotten their degrees—and "sticks"—hurdles to be overcome at every step along the way, and dismissal for those who fell behind—paced the aspiring scholar. Thus, while the American graduate school perpetuated the German university's dual commitment to education and research, there was a major difference in the institutional manner in which this commitment was carried out.

The modern German university was not originally built for the natural sciences. Its creativity was first expressed in the classical studies and philology of the early 19th century; it was in these fields that the seminar system originated. Only after 1830 did the center of gravity shift to the natural sciences. Thus the university was chiefly a place for listening to lectures which students were left to choose for themselves. There was no curriculum or graduation. Under these circumstances, the establishment of a graduate school would have been meaningless, for students were free to show up in any classroom regardless of age, year in study, or qualifications.

In the modern natural sciences and engineering, normal science traditions had made such solid and extensive advances that coherent basic training could not be adequately carried out where students were given free rein to select their own courses. No longer able to keep pace with the rapid pace of progress that resulted from the normal science research it had initiated, the German university lost its position as the most advanced institution of its kind. In a *laissez-faire* system, it was not possible to carry out laboratory instruction and other kinds of training that required the planned use of facilities.

Beginning as it did with the needs of the natural sciences, the American graduate school emphasized training at the expense of free-ranging intellectual inquiry, devising an efficient system for drilling prescribed sets of normal science procedures into the heads of students and budding scholars. Under this system a new type of scholar was born and a new type of academic tradition came into being. Its main features included:

1) A willingness to broaden the borders of academia. Having gotten its start in a new age, it conferred degrees in things like physical education, the arts, and even home economics—things that were, from the point of view of the German *Wissenschaft*, not academic disciplines.
2) Production of scholars with a strong professional consciousness. With formal education continuing until a relatively advanced age, graduate students inevitably became absorbed in their special fields. The wide-ranging speculative thinking that one sometimes still comes across in European students was expunged, and academics became less intellectuals than specialists. (It should be noted, however, that the system was designed so as to encourage familiarity with as many neighboring or related fields as possible in order to prevent excessively minute specialization and to promote pioneering work in border areas.)
3) Production in large numbers of highly effective normal scientists.
4) Complete professionalization of research and researcher. Research itself became more responsive to an artificially established merit system and better suited to funded, project-centered work, while the status of the scientist underwent a change that facilitated the scientist's entry into the worlds of business and government. With its classical freedom of research, the German university had produced scholars who engaged in academic science, but the American graduate school brought forth professional researchers who believed in efficiency and proved adaptable to industrialized science.

One could note other special features. In general, however, one can say that the American graduate school laid the groundwork for a fundamental change in the nature of science, from a scholarly activity pursued by a small number of freethinking, gifted individuals to one that often succeeds by the sheer depth of its B and C class of technically competent researchers. This was the change from 19th-century to 20th-century science, the transmutation of science into big science.

Neither the consequences nor the potential significance of the American tradition was immediately apparent, even to Americans. The United States continued to feel inferior to Europe. The quality of its teachers and the lower degree of popular appreciation of scientific research seemed to confirm the feeling. Study in Germany remained the conventional way for the young scholar to enhance his stature. American scholars in science continued to regard Germany as their spiritual homeland until after World War I. In this respect the interwar period proved to be a watershed. After the First World War American science began a movement toward the highest world standards that led to its present supremacy.

If the Soviet Union now challenges this supremacy, the Soviet higher educational system can perhaps be described as the American graduate school further reinforced as a training organization. In the Soviet system the number of researchers produced can be more precisely controlled than in the United States. There is a systematically designed, artificially constructed merit system, and government policy assures that the natural sciences and technology enjoy a position of advantage in the academic world.

Scholarship Moves Outside the University

In the 19th century the university was the center of scholarly research, but educational institutions today have no exclusive claim on the task of research and development. That in the 20th century universities have come to share the research spotlight with government and business institutions is no more alarming than the fact that at the end of the 18th century universities themselves replaced the academies as the driving force of scientific development. For one thing, by the 20th century the university had begun to produce scientific technicians and researchers beyond the capacity of higher educational institutions to absorb them. This surplus moved into the worlds of government and business, to become a professional group with its own place in society. Behind this outward movement, there was of course recognition by both government and business of the need for science and technology and of the objective conditions in which they flourish. At least during the early stages, however, this awareness was the work of professional scientists themselves, who had gone through a kind of modern training unknown to the intellectual or craftsman of the past. Moreover, the initial contacts that gave rise to this awareness were due less to a conscious attempt on the part of government and business to bring scientists into their organizations than to the efforts of scientists to sell themselves by extolling the practical virtues of the scientific method.

Having made initial inroads into government offices and business enterprises, scientists sought to strengthen their position by bringing in others of their own generation and the next. These scientists not only found a new type of employment; they also generated a form of scholarship different from that in the university. Though some important research was conducted outside the university in the 19th century, non-university research really came into its own in the 20th. Today any discussion of scientific research must include what goes on in three different types of institutions—the university, government, and business and industry.

Government and Science

The history of government involvement with scholars and scholarship is a long one, but in former times it was often more sporadic than sustained. It typically took the form of personal patronage by the monarch or ruler, as we have observed in the Islamic world and the Renaissance states. When a monarch died or was deposed, the scholarly tradition that had prospered under his favor ordinarily declined. The only exceptions to this general pattern one can think of are Chinese calendrical astronomy and Confucianism, both of which were securely rooted in the bureaucratic system. Thus, until the 19th century, beneficient patronage of learning was more likely to be found in absolute monarchies than in democratic polities. Patronage of science was in fact so closely identified with monarchical governments by the early 19th century that it received little support from the U.S. Congress, which saw it as undemocratic.

The French Revolutionary Government was the first to recognize that the protection and encouragement of science and learning was vital to the modern state and to make it a part of policy and institutions. In the 19th century, this way of thinking began to take root in all modern states, but science and technology policy itself remained at the formative stage. The systematic policies that we are familiar with today did not emerge until the First—or, in many cases, the Second—World War.

War has traditionally been the most important activity of the state, and military technology one of the major concerns of rulers. Throughout much of the 19th century, however, military weaponry was largely a matter of "how many" or "how much"—that is, a question of the size of the budget. Little thought was given to producing revolutionary new weapons through investment in scientific research. Yet by the last third of the century, army technicians trained in military schools were no longer able to make the rapid progress in military technology that governments had come to desire, and technical specialists who had had the benefit of a regular modern engineering education began to take over the field. Ordinary scientists, however, did not become deeply involved in military research until the mobilizations of the World Wars.

The modern state has many functions: setting up and operating an effective educational system, providing health care, constructing and maintaining public works, legislating and administering regulations protecting industrial ownership. Finally, modern governments engage in academic surveys and research. The Coast and Geodetic Survey established by the U.S. government in the 19th century represents the type of activity that is more efficiently accomplished through large-scale, well-funded government projects than by single individuals in university research laboratories. As the American frontier developed, the government also conducted land surveys. The United States pioneered in field sciences such as astronomy, meteorology, and geophysics. These were largely peripheral to the laboratory-research-centered science of the German university, whose major contributions lay in chemistry and physiology.

As industry advanced, governments also responded to the need for precision measurement, inspection of machinery, and testing of materials. The German Physikalisch Technische Reichsanstalt was established in 1887, and other nations followed suit—England with its National Physical Laboratory, the U.S. with its Bureau of Standards, and Russia with its Mendeleef Institute. In its Kaiser Wilhelm Society (now the Max Planck Institute), Germany was also the first to create the kind of modern national research establishment that linked scientific research and development with national prestige and hence national policy.

Business and Research

Watt's steam engine symbolized the technology of the English industrial revolution—traditional artisanry which did not as yet have much contact with science. With the coming of the 19th century, however, the situation began to change. After 1820, business and industry gradually began to recognize that capital invested in experiments and trial manufacture might lead to greater profits, and scientific research methods began to be applied to technological development. Again it was Germany and the United States that were to lead the way—a phenomenon that can be at least partially explained by the fact that the modern industrial sciences (chemistry, electricity, etc.) came into their own just at the time of their industrial revolutions so that links between the two were established at a relatively early stage.

The iron and steel magnate Alfred Krupp created a laboratory for analytical research in 1862 and a materials testing laboratory at about the same time. By the 20th century, the Krupp experimental and research facilities for the study of iron surpassed anything to be found in the universities. Elsewhere, after a series of scientific experiments, inorganic chemists succeeded in developing technology for the industrial manufacture of soda by the Solvay process. In the world of organic chemistry, extension of small-scale laboratory methods produced industrial processes typical of the new, industry-related scientific technology. University chemists had established connections with industry as early as the first half of the 19th century, and the achievements of men like Justus von Liebig and Sir William Henry Perkins are well known.

Finally, there was the work of Bell and Edison, as a result of which electromagnetic phenomena that had only been subjects of scientific research during the first half of the 19th century suddenly became the nucleus of a new electrical industry.

During this period, science and engineering specialists with university-level training began to move out into the industrial world. In the 1870s and 1880s business employed full-time researchers for the first time, and the technological development that had heretofore been left to amateur inventors and their ideas became a large-scale enterprise. By the end of the century the largest German enterprises had research staffs of several hundred persons, while in the United States, the early 20th century witnessed the creation of the famous General Electric and Bell Laboratories.

In the interwar period, rivalry between university, government, and research institutes became a feature of scientific research in capitalist countries, giving rise to a new classic formulation of the differences between them: the university has freedom but no money, business has money but no freedom, and government research has a little of both.

As the formula suggests, the researcher in a non-university research institute is freed from educational tasks but finds himself ensnared by the bureaucraticism of government and the profiteering of private industry. Values extraneous to the purposes of science and learning take priority. The university belongs to the scholar, but in government and industry he is a second-class citizen, subordinate to high-level bureaucrats or managers. This does not necessarily mean that the scholar in government or industry is poorly treated, but he is experiencing something foreign to the professional ethic of academic science: the subordination of originality to other values. In one survey, American scientists in business and government expressed more dissatisfaction with their jobs than did management personnel (Kornhauser, 1962). While many scientists said that if they had it to do over again they would prefer managerial jobs, few managers expressed a desire to become scientists.

In 1918 Thorstein Veblen was already deriding the introduction of the entrepreneurial principle into the university and the decline of that 19th-century atmosphere in which the search for knowledge was the supreme imperative. A university dominated by the principle that salesmanship is more important than workmanship could carry out routine work, but, he admonished, it would do nothing more (Veblen, 1918). Veblen's forecast has grown more and more true with the passage of time, prompting at least one critic to speak of contemporary science as "industrialized science" (Ravetz, 1971).

If one thinks of research output as like factory production, then the research institute may well be more efficient than the university. Mass production is, after all, more efficient than individual craftsmen making one article at a time. But this is true only where the assignment is the speedy resolution of ordinary normal-science problems. And there are other issues. Who is to define the problems? Who will build the paradigms? Who will develop and shape them into a scholarly tradition? To ponder these questions is to become aware once again of the significance of maintaining ties between research and education. Unless infused with new blood, scholarship at research institutions continues to revolve around the problems and perspectives of a single generation. Formidable minds grow weary of their work, forget the significance they once attached to the puzzles that lay before them, and lose their intellectual vitality. Herein lay the weakness of the Kaiser Wilhelm Institute. In the absence of contact with and the challenge of students, its scholars could not avoid a growing conservatism and a narrowness of political vision, and they eventually found themselves caught in Hitler's web.

Big Science

The emergence of the German university as a center of scientific advance resulted in an increase in the number of scientists. The pioneering Japanese physicist Nagaoka Hantarō, who studied in Germany from 1893 to 1896 and visited Europe many times thereafter, once observed that while France and England produced a small number of exceptional scholars, German scientists ran the gamut from the first-rate down to the very mediocre. This depth was a product of the 19th-century German university. Half of what is said about German scientists being steady, systematic, and sometimes even a bit pretentious can be explained in terms of an institutional setting that led the way in providing conditions for the professional employment of scientists and was the first to produce them in large numbers.

This mass-production approach was taken over by the United States in the 20th century. The Japanese scientific community paid no attention at all to American science until after the First World War, and found it of little import until after the end of the Second World War. Compared to Europe it was considered second-class. But what was the U.S. doing all this time? After the Civil War, proceeds from the sale of federal lands were made available for the establishment throughout the nation of higher-education institutions with applied curricula. With the coming of the 20th century, American graduate school education began an explosive growth that has only recently leveled off. These institutions did not produce the top-notch scientists to be found in Europe, but they did turn out large numbers of second- and third-level scientists and technicians. Taking paradigms that had originated in Europe, they turned out normal science work in great quantity.

The growth of science education was underwritten by a thriving economy. The median living standard in the United States had already surpassed that in Europe in the 19th century, and there were plenty of young American scientists willing to pay whatever it might cost to study in Europe—particularly Germany. Blessed with an abundance of facilities, America built giant telescopes and advanced the study of astrophysics and space. The nation also took in many fleeing German scientists, some of whom, working in American facilities, helped to produce the atomic bomb.

The World War II mobilization of scientists led to the destruction of the three-cornered balance between university, government, and business research institutes, and greatly intensified government involvement in the world of science. Since the war, the Soviet Union has also produced large numbers of scientists and invested heavily in research facilities.

As scientific research came to be dominated by massive projects involving large numbers of researchers and large sums of money, science underwent a change that has affected its very constitution. Today scientific research has been transformed from an operation carried on in university laboratories by single individuals or small groups to an enterprise dominated by large research institutes where numbers of researchers and organization of work are the decisive factors.

Let us grant that a research institute is the most efficient type of organization for normal science research. But let us also note that it is no more the creature of a voluntary paradigm support group than any other large-scale organization. The paradigmatic seat is occupied not by a Copernicus or a Newton but by a government minister or company executive, and the researcher pays his respects not to heliocentric theory or Newtonian mechanics but to a money-waving administrator who commands: "Make missiles" or "Go to the moon." If what is operative here can be called a paradigm at all, it is clearly not the kind of idea paradigm that emerges in the normal course of scholarly development. It can only be called an institutional paradigm, or perhaps a money and power paradigm.

Let us suppose that a 19th-century scientist were to compete with a 20th-century one in attempting to calculate the orbit of an artificial satellite. Since both would base their calculations on Newtonian principles, the 19th-century scientist would not be at a particular theoretical disadvantage. But even if he worked all night, night after night, with a table of logarithms at hand, he would find the extensive calculations very hard to handle. By organizing a research team and making use of the computer, however, the scientist of today could have the answer in short order. The difference—the decisive difference—is one of material resources and organization. If we are to understand why 20th-century scientists can do things that 19th-century scientists could not, it is these two factors with which we must deal. Contemporary science is "big science," and big science is the result of the convergence of material resources and systematic organization.

Not many decades ago, when a small country like Denmark was blessed with a paradigmatic scientist like Niels Bohr and his Copenhagen group, a position of world leadership in science was by no means beyond its reach. Today, however, the human and material resources that have been mobilized by the United States and the Soviet Union give them a marked superiority. When manpower and money can get scientific research done, even the emergence of a creative genius is of little avail. This is the nature of the scientific community in our time.

Big science is a method of accelerating the cumulative progress of normal science in a predetermined direction. Perhaps it would be more appropriate to call it a means. A scientist working on a giant project in this stream of accelerated progress may be rewarded by the feeling that he is personally involved in the "evolution" of modern society. But have not scientifically revolutionary efforts always fought against the stream—worked to change its course? The classical proposition that scholarly activity ultimately springs from the innovative ideas of individuals is still true today and will continue to be so. So long as scholars continue to be individualistic, the tension between organization and individual research is bound to be even greater in the future.

In J. D. Bernal's first work, *The World, The Flesh and the Devil*, (1929) the English systems thinker posed many of the classic questions of futurology and forecast several of the problems of big science today. The idyllic age in which scientists approached nature like primitive people foraging through field and mountain was, he observed, a thing of the past, by virtue of the same logic that led to the recognition that cultivation was clearly a more effective way of obtaining food than picking up fruit off the ground. Hereafter science would move more and more toward organized research. Bernal perceived the dangers that would accompany this development and pointed to the need for educational and institutional reforms that would enable each individual researcher within a research organization to reconcile his creative ideas with routine work. Today big science has become a reality, but the reforms necessary to sustain the individual researcher in its midst are yet to come. This is the source of many of the conflicts that plague most research groups. Is it not high time that sociological insights be applied to the mechanism of big science?

Historically there has never been an instance in which a revolutionary idea destructive of the existing scholarly framework emerged at the behest of an organized structure. As I noted in Chapter 1, debates and discussions in small groups generate ideas; but large groups require administrators, who either suppress ideas or sit by and watch while the group buries them.

In the nature of things, organizational structures properly succeed ideas. As I indicated in Chapter 2, they are in essence practical responses to the needs of paradigm support groups (initially the free associations of people committed to the same paradigmatic idea) at a particular stage in their evolution. An amateur scientist of the 19th century could simply give up scholarship when he ran out of ideas and lost interest, but today's professional research groups and their individual members remain in business even after their creative energy has been spent, protecting their organization and, in the process, almost inevitably working to constrain the ideas of newer groups.

If vitality is to be maintained, the classical pattern of scholarly development—generation of paradigmatic ideas—the formation of a paradigm support group—the development of a normal science program—must be preserved within the organizational structures to which scholars are professionally attached. Moreover, these structures must be constituted so as to permit a constant gathering and dispersal of researchers around idea paradigms in response to individual projects. In other words, big science should be concerned less with big money and big organization than with big ideas. Little is to be expected from scientists who feel more allegiance to organizational structures than to idea paradigms.

Critical Science

Thus far I have discussed a variety of paradigms and their development. In doing so, the most basic paradigm of all, the paradigm of scholarly activity and advance itself, has been accepted without reservation. My intention has been to place more emphasis on revolutionary scholarship than on normal scientific progress. I have taken this approach because with the protection and patronage of the power structure, normal science faces no impediments to progress. The word "progress" is now trite; the idea that modern science and technology lie at the heart of this progress is entrenched in modern thought. Even practitioners of other disciplines such as the social sciences have been persuaded that by imitating the ways of modern science they too can climb aboard the bandwagon of cumulative advance.

Modern research groups have come to be solidly organized, and have seen their expansion and reproduction guaranteed through the modern educational system. Moreover, in recent years the protection and promotion of science has become a vital part of the cultural and industrial policies of the modern state. A constitutionally anti-science regime, like that of the Nazis, which tries to suppress or pervert the development of universalistic science, is not likely to appear again, for it has been demonstrated that such regimes only dig their own graves. Inasmuch as science and technology are part of the very fabric of human society, mankind can be expected to continue its headlong dash down the open road of progress.

But where will all this progress lead us? People have pondered this question at least since the time of *Gulliver's Travels*. Until a generation or so ago, however, criticism of modern science—and, more broadly, of the scholarship that took modern science as its model—came chiefly from poets and religious thinkers, from the humanistic flank. Most of it was part of what is customarily referred to as the "Romantic reaction," which insisted that not everything in human life could be explained by modern science. Since the invention of the atomic bomb, however, science has become an unprecedentedly serious social issue, the subject of parliamentary controversy and international political dispute. Many people have come to feel that the unthinking development of science will eventually sweep us all away and destroy us. If the end of scientific progress is atomic destruction, is it not time to go back and re-examine the mechanistic paradigm from which modern normal science took its departure?

To suggest that there is a problem with the paradigm to which scientists have committed themselves, and that if the paradigm were changed, traditions of normal science might be given a new direction, is to call for a scientific revolution. We must pause, however, to ask whether a critique of science directed at the scientific method or system in the abstract can be effective. People once opposed machines for many of the same reasons. Machine civilization, it was urged, meant the mechanization of human life, the reduction of human beings to cogs in a wheel, the bridling of human liberty. "Machines are the enemies of humanity," went the cry; "cast them aside and return to truly human life!"

But machines themselves do not sin. When machines become the object of resentment, they provide cover for the capitalist, the industrial producers, the beneficiaries of "machine civilization," turning aside criticism that would otherwise fall upon the humans behind the machines. The machine is, in other words, scapegoat. Does not the recent anti-science mood have its eyes fixed on much the same kind of "imaginary enemy"? Science itself is not evil. The problem lies with the scientists who practice it and the way in which they have come to identify themselves with established structures.

After the Bolshevik Revolution, progressive scientists in many parts of the world anticipated the emergence of a new, ideal, Soviet science, uncontaminated by capitalist motivations, and there were a variety of attempts within the Soviet Union to respond to new expectations. In the end, however, the results were disappointing.

The new China has made scholars rather than science the issue and set out to create a new breed of scientist. The "barefoot doctors" who received so much attention during the Cultural Revolution, and the periodic return of scholars and students to the countryside, are among the more dramatic expressions of this orientation. After long centuries of the civil service examination system, this radical approach was imperative if the image of the scholar and scholarship as part and parcel of the power structure was to be overturned. Although the Chinese experiment was conducted under unique social conditions, Chinese policy attacked the right problem—scientists rather than science.

The professional behavior of doctors may be a legitimate target of criticism, but few people are opposed to medical treatment. Similarly, opposition to science should first take a critical look at the scientist himself. Have scientific groups become too comfortably settled in the system and too subservient to government and industry? Perhaps what is needed is more critical scientists who will exercise some judgment over and give some direction to the present blind advance.

In outlining his own hopes for the rise of "critical science," Jerome Ravetz (1971) has distinguished between modern "academic science" and the "industrialized science" of today. By "academic science" he means something like what went on in the pre-industrial German university—the neither useful nor harmful science that I have described in the first half of this chapter. The term "industrialized science" refers to the science I have been describing— a science that manifests itself in powerful technologies, and that is managed just like a government office or business or in line with their interests and desires. But we cannot simply return to "academic science." The university academism that supported it has been decimated by economic growth and lacks its former vitality. If we are to correct the flaws and corruption that the "science industry" has spawned we must, with Ravetz, think in terms of the emergence of a new, critical science.

The contours of this critical science are as yet unclear. It may well be more argumentative than analytical, treating people and things together rather than simply dealing with things in isolation. Yet, as I have suggested, the appearance of a group of critical scientists attended by a revolution in the system of values which science serves would be of greater significance than changes in the method or substance of science itself.

The contradictions of big, industrialized science and the disaffection one hears at every turn suggest that we are now in a pre-revolutionary situation. But where is a scientific revolution to break out? Certainly it will not be in the giant institutions of big science. Perhaps it will come from citizens' movements, or from oppressed groups such as blacks or women. Again it might emerge from the developing countries.[5] Any of these groups might conceivably develop a science distinctly different from that which now exists.

In times past, criticism of science was rooted in upper-class aversion to the industrialized sciences of the upwardly mobile bourgeoisie. It came chiefly from the aristocratic and landlord establishments. At the present time, however, when the old upper classes have lost both their creative powers and their political voice, and science has become part of the establishment, critiques

[5] At some point it will be necessary to think about critical science as more than a short-term movement and make some kind of institutional arrangements that will ensure its existence on a permanent basis. It may be suggested that the new United Nations University and its research institutes should be oriented toward this kind of critical science.

of science are coming from the forgotten children of technocratic culture, from students, young scientists, technicians. Resentment of industrialized science as it now exists is also evident in developing countries as they watch the developed countries use their monopolistic control of science and technology to widen the global imbalance between rich and poor. There is no reason to believe that laymen and private citizens will not continue to take up the issues of big science. Can a new paradigm emerge from the midst of these discussions and movements, survive its critics, and lead to the creation of a new line of scientific research?

Service Science

From the viewpoint of my version of the sociology of science, which emphasizes the role of the receivers or assessors rather than the producers of scientific achievements, I have suggested a new concept of "service science," which is defined as science assessed by people at large, in contrast to the academic science assessed by fellow peer group members and industrialized science assessed by public and private sponsors (Nakayama, 1981). Service science may be defined as science sustained by a sense of solidarity with and sympathy for local residents and responsible to their standards.

The first characteristic of service science is that it does its research in the field, working to solve problems like environmental pollution that are of concern to the public and not easily accessible to laboratory investigation alone.

Service science requires forms of publication that differ from the scientific paper or the report. It addresses its findings not to fellow scholars or to research project administrators but directly to local residents.

Inasmuch as service science is closely tied to particular local areas or regions, it would not be at home in elite universities or research institutes of business enterprises. Local government research institutes and local institutions of higher education may be the ideal means of institutional support for service science.

Special education and training would be necessary for practitioners of service science. More effort would be made to harness social consciousness to scientific education, to involve students and members of the community in selecting research projects and carrying out manpower-intensive research.

The concept of service science may seem naive by the standards of academic science or industrialized science. But we are talking about a new way of thinking about science problem-solving that cannot meaningfully be evaluated by the standards of the academy or industry. There should be a place for a service-oriented type of science alongside the more traditional types.

In the Western concept, the "professions" of theology, law, and medicine have, since their origin in the medieval university, been identified with a group of intellectuals who have university training and give at least normatively intellectual service to the people, independent of political and economic pressure and interest. In contrast to these older professions, the scientific profession lacks a sense of the public as its clientele, and this lack is enhanced by the fact that scientists are usually supported by public or private authorities. I am trying to expose this lack of a sense of service toward the people in present scientific enterprise by coining the new term "service science," which may function as a "shadow paradigm" as long as scientific activity remains alienated from ordinary citizens' lives.

Chapter 6

The Transplantation of Modern Science to Japan

What happens to a paradigm and the scholarly tradition that has grown up around it in one culture when it is introduced into another? Among people who employ a different scholarly jargon and work under different social conditions, the paradigm confronts new challenges that occasionally affect its very structure as a mode of learning. The removal of Greek learning to the Islamic world, the introduction of Chinese culture into Nara–Heian Japan in the 9th and 10th centuries—history does not want for examples of this phenomenon. Here I would like to take a look at what is perhaps the most dramatic instance of all, the entry of modern science (that most excellent product of the modern West) into a non-Western country—Japan—for the first time. More was involved in this case than the gradual spread of a cultural achievement from one region to another. The influx of modern science into Japan was intimately bound up with the policy decisions of a modern state.

A scholarly tradition that emerges from within normally follows the following course: 1) generation of the paradigm; 2) elaboration through the formation of disciple-advocate-support groups; 3) canonization (incorporation into textbooks or manuals); 4) institutionalization. The level on this spectrum at which a foreign paradigm is transplanted determines its form of acculturation.

Inasmuch as the academic traditions of East and West have grown out of different paradigms and present different spectrums, the reception of Western scientific thought in Japan could not have taken place at the paradigm level. The paradigms of modern science, in other words, could not have been subsumed under traditional paradigms. This had already proved impossible even in the West—indeed this is why there was a Scientific Revolution;

and even if scientific paradigms had been successfully assimilated under traditional paradigms, they would, by definition, have ceased to merit the designation paradigms. The introduction of the basic paradigms of modern science involved the transformation of old concepts and the creation of new ones. When Shizuki Tadao first introduced Newton to East Asia in the late 18th century with his translation of John Keill's (1701) canonical interpretation of the Newtonian paradigm (*Introductio ad veram physicam*, translated as *Rekishō Shinsho*), there were no words in the Sino-Japanese vocabulary that could adequately express Newton's notion of particles or his concept of mechanics. In attempting to ground Newton in the East Asian concepts of *i* (change) and *ch'i* (matter-energy), Shizuki may well have been taking a first step in the direction of a new natural philosophy, but in the absence of a mature support group, his work was not expanded into a new scholarly tradition.

In the contemporary world, where people in professional disciplines like physics share internationally accepted goals that transcend particular cultures, a paradigm that appears in one country quickly finds supporters in many countries, who have the capacity to put the paradigm to work in a normal science program. In this sense, the international physics community today may be said to share a common culture, and the transmission of a paradigm follows the course outlined above.

But when a paradigm is transplanted from another culture, so that there are no established support groups sharing common philosophies or outlooks to receive it (and in the case of modern science this precludes the existence of a normalized research tradition), the sequence is rather different. In the absence of participation and feedback in the paradigm-elaboration process, the paradigm brought in from abroad is treated as an established canon to be faithfully translated. This pattern is commonly observed where contact with a foreign culture has led to a gradual osmotic acceptance of its cultural achievements by various individuals in the private sector, as exemplified in Japanese history by the ancient arrival of the Buddhist classics and the advent of "Dutch learning" (Western scholarship that was introduced via a small Dutch settlement) in the late 18th to mid-19th centuries.

There is also another pattern. When foreign learning is systematically introduced by the state, the establishment of an educational system under government auspices comes first. Paradigms and canons become part of the school curriculum and advocate-support groups are created as a part of national policy. Though this has become typical wherever scholarly paradigms have been imported on a national scale by modern states, Japan's introduction of Western science after the Meiji Restoration (1868) remains the classic and earliest example of this pattern, and this will be examined later in this chapter.

Western Influence on Traditional Sciences

But first let us turn to the earliest influence of Western science in Japan. In the mid-16th century Jesuit evangelists "discovered" Japan; soon after, however, the Japanese government, considering the influence of Christianity detrimental to the cohesiveness of their culture, banned all foreigners from the country, sanctioning trade only with the Dutch through the port of Nagasaki, off the island of Kyushu. This relatively effective, self-imposed containment remained in effect until the mid-19th century. Western ideas did, nevertheless, seep through, directly and indirectly.

There were at this time (the Tokugawa period, 17th to mid-19th centuries) four traditional sciences, each with its own paradigm and professional groups: 1) calendrical astronomy, 2) medicine, 3) materia medica, and 4) mathematics (*wasan*). A spectral analysis of these fields reveals that Western learning was accepted somewhat differently in each case, depending upon the nature of the science, the character of its practitioners, and the strengths and weaknesses of the native tradition and its classics.

1) *Astronomy*. In astronomy the influx of the new Western paradigm did not spell destruction of the old. Data generated by the new were simply incorporated into the old framework.

Traditional astronomy was Chinese-style astronomy for calendrical calculations. Its purpose was to investigate the apparent movement of the sun and moon, to put them together to construct the luni-solar calendar, and to predict solar and lunar eclipses. The many systems for computing the ephemerides incorporated in the *Lu-li-chih* (chapter on the calendar) of official Chinese dynastic histories served as the paradigms for this astronomy. The Yuan calendar (*Shou-shih li*), which was considered to be the crowning achievement of the traditional Chinese astronomy, was particularly important. Its mastery was the final step in obtaining a license in traditional mathematics (*wasan*).

Western astronomy had arrived in China with the Jesuits. Calendar-making was a vital function of the Chinese bureaucracy, and the Jesuits hoped that by demonstrating the superiority of Western astronomy they would succeed in persuading the bureaucratic elite that Western culture and Christianity were superior. Astronomy was well suited to their purposes, not only because it was so important to the Chinese bureaucracy but because the phenomena with which it dealt possessed a universality that transcended East and West and the objectivity of celestial phenomena did not permit of much human manipulation. Furthermore, it was quantitatively precise. Thus, when the tally sheet was in, Chinese astronomers had to acknowledge that Western astronomy had superior features. All that the Jesuits made available from Western tradition, however, were peripheral data and methods of calculation. The structure, style, and purposes of Chinese calendrical astronomy remained unchanged. As Hsü Kuang-ch'i, a high Chinese official who collaborated with the Jesuit Matteo Ricci on several projects, put it, "We melted down their materials and poured them into the *Ta-t'ung* (the traditional Chinese calendar) mold."

In Japan, the eighth Tokugawa Shogun, Yoshimune (1684–1751), attempted to use this refurbished Chinese astronomy as the basis for calendrical reform, but he was effectively opposed by conservative court circles in Kyoto. The Hōreki revision that came into effect in 1754 was still modeled upon the traditional *Shou-shih* calendar. Yoshimune's desire was fully realized, however, in the Kansei revision of 1797.

The Tempo revision of 1842 made use of Western astronomy learned directly from Dutch books. Yet even here the framework of Chinese-style astronomy was not broken. Japanese calendar-makers, like their Chinese predecessors, were chiefly interested in the more precise numerical data of Western astronomy. Someone like Takahashi Yoshitoki (1785–1829), who worked in the tradition of Asada Gōryū, could dabble in questions of planetary motion out of intellectual interest, but the subject was peripheral to the official task of calendar-making. Indeed, the new Western paradigms of modern astronomy—theories of the cosmos and celestial mechanics—had nothing to contribute to traditional paradigms. No attempt was made to fit them into the Chinese mold.

Calendrical astronomy was official learning and government-managed science. When Western learning entered the country it initially became the work of the official astronomers of the Tokugawa Shogunate. With the fall of the Tokugawa regime in 1868, the office and the routine observational work its occupants had conducted were abolished. Even if it had survived it is unlikely that the Tokugawa astronomers would have become modern scientists. A remnant worked on the solar calendar reform that came into effect in 1872, but given their training this routine computation was about as far as they could have gone. Thus the field of astronomy passed into the hands of modern professionals, newly graduated from the university. Between the two generations there were no teacher-disciple relationships, no continuity. In summary, the traditional paradigm in this field proved so strong that it kept the incoming paradigms of modern science at arm's length, and Japanese astronomers made only supplementary use of the data they generated, using them to strengthen the fabric of their tradition.

2) *Medicine.* In medicine, Chinese and Western paradigms coexisted peacefully, for there was little overlap in their areas of strength and hence little room for competition.

Western medicine first came to Japan as the *namban igaku* (medicine of Southern barbarians) that entered the country with Christianity in the 16th century. Before long, "barbarian surgery" was being welcomed to compensate for deficiencies in Chinese medicine. Yet what was picked up at this point was not an organized body of knowledge but rather techniques such as rubbing salve on wounds and the use of alcohol as a disinfectant.

When we say that science is governed by paradigms, we mean that paradigms determine the shape of the questions posed and the answers to be delivered by normal science. But there are some fields in which their dominion is far from total. This is particularly true in medicine, where the mechanistic paradigms of modern science were not readily applicable to the many intricate and varied techniques used in dealing with the human body and all its complexities. Older paradigms peculiar to medicine and biology have continued to have meaning. The relationship between theory and empirical techniques was even more diffuse in Chinese medicine where physiological and pathological theories were based on doctrines of *Ch'i*, drawn from Yin-yang and the Five Elements natural philosophy. These theories were indeed little more than window-dressing in Japan: they were, after all, imported from China, and much less coherence between theory and practice was maintained than in China. Medical techniques governed by *Ch'i* theory, however, have survived on their own down to the present day, without the support of the old paradigm.

Scientific medicine, in contrast to bodies of medical techniques accumulated through experience, approaches treatment on the basis of physiological and pathological theories that are grounded in anatomy. Chinese medicine also had its theories of physiology and pathology, but anatomy was considered to be rather unimportant: the source of disease was not traced to a particular organ but attributed to the disharmony of the *Ch'i* permeating and circulating in the body. To the comparative eye this appears even weaker after the European Renaissance, when great strides were made in anatomical studies, and the 17th century, when contact with mechanistic ideas opened the way to modern physiology with its attempts to explain the functioning of the human body in mechanical terms. Seventeenth-century Europe also witnessed pioneering work in the chemical treatment of disease.

Nevertheless, it is extremely doubtful whether this kind of basic research contributed much to actual medical care before the 19th century. Half-baked theory could on occasion even lead treatment astray and, once embraced, serve to deprecate and discourage the use of "popular" empirical remedies. All medical theory of the time regarded the quinine treatments of the Peruvian Indians as popular superstition, but there were doubtless many patients who were thankful for them. Much medical theory was little more than academic trapping, and one may say that the paradigms of modern science had yet to become accepted as a definitive guide to practical medical care.

Under the circumstances attempts to argue for the superiority of Western medicine were not very persuasive. Western superiority might be accepted in surgery—undertaken as a last resort, since at the time it often proved fatal—and such demonstrably effective procedures as vaccination, but these were exceptions. Insofar as general internal medicine was concerned, there seemed to be little to choose between China and the West. Since, regardless of theory, whatever worked was best for the doctor as well as for the patient, the average Japanese physician adopted a syncretic approach, treating patients with a combination of Chinese and Dutch medicine.

Still, the theoretical paradigms of the two traditions were scarcely compatible. It is at this level—and not because of any contribution it made toward advances in medical treatment—that the appearance of the *Kaitai Shinsho* (New Treatise on Anatomy) can be considered an epoch-making event. This 1774 translation of a Dutch version of a German work on anatomy represented the first real evidence upon which the advocate of Western medicine tried to make the case that Western medicine's much greater knowledge of anatomy meant that its methods of treatment must also be superior. Although anatomy was not relevant to actual medical treatment, it gave Japanese physicians an opportunity to become acquainted with the anatomical basis of the localized approach to treatment that, after about 1800, was the distinctive feature of Western medicine. In East Asian medicine, physiology and pathology rested on the operations of a *Ch'i* that filled the universe and the human body, and medical treatment centered around assisting the whole body to return to a state of harmony within and without and recover its normal functions. Early 19th-century Western medicine, however, looked at the body as a machine and saw pathology in solidist terms. The problem was to locate the part of the body that was ill and treat it in a concentrated fashion.

In a scientific revolution the revolutionary process ordinarily reaches completion when the new paradigm succeeds in encompassing the empirical knowledge that was explained in a different way under the old paradigm. At least this will be the objective. But modern, chemotherapy-centered medicine has found Chinese herbal medicine and techniques such as acupuncture and moxibustion difficult to cope with. In China, Chinese and Western medicine continue to coexist and are thought of as complementary—but are integrated only at the level of treatment. In the Japanese case, the creation of a modern medical system occasioned a political struggle between Chinese and Western support groups. The older physicians who were the chief backers of Chinese medicine were defeated. They were excluded from the new licensing system of 1876, after a century of attack by spokesmen for "Western" medicine, and the medicine they practiced came to be regarded simply as a bag of popular remedies.

3) *Materia medica.* The encounter of East and West in this field led to the growth of a new branch of normal science through the addition of empirical data from the East to the Western paradigm or vice versa.

In the Chinese tradition, there was a field of study known as *pen-ts'ao*. This field was mainly concerned with the medicines used by Chinese medical practitioners, but it also included identification of animals and plants mentioned in the Chinese classics with local Japanese flora and fauna, and "product study" which compiled information about useful animals, plants, and minerals found in various parts of the country. It was chiefly concerned with collection, classification, and description, and developed an artificial-classification system that appears in the *Pen-ts'ao kang-mu*. This system was based on use and external appearances, and distinguished mountain herbs, tropical herbs, poisonous herbs, medicinal herbs, fragrant plants, trees, and shrubs.

Japan was introduced to the natural classification system of Linnaeus in 1829, primarily through the work of Itō Keisuke. The *pen-ts'ao* and Linnaean paradigms had little to say to each other, but Itō had studied both, and by explicating Linnaeus' classificatory scheme and identifying Latin names of Japanese plants, he made possible the addition of Japanese data to the Western mode of natural history. Itō's methods, however, differed little from the collecting, classification, and description of the traditional Chinese herbologist. He does not seem to have understood the modern method of examining structure, properties, and patterns of growth. Though his contributions were acknowledged and he lived to see the dawn of the 20th century, he became increasingly out of touch with students trained after the 1870s. Since Chinese herbology was traditionally carried out mainly by physicians as an adjunct to medical practice, and had no independent group of professional practitioners comparable to the pharmacists of the West, it was destined to share the same fate as Chinese medicine.

4) *Mathematics.* In mathematics, Japanese and Western paradigms were self-contained systems, mutually exclusive in structure and constitution. When the Meiji political revolution and institutional change came, there was a complete change from one to the other.

Wasan (Japanese mathematics) had roots in the Chinese mathematical tradition, but it represented an independent development. Taking the work of Seki Takakazu (?-1708) and Takebe Katahiro (1664–1739) as paradigmatic, it emphasized the algebraic solution of geometric problems. Although this development involved a turning away from practical concerns in favor of mathematical puzzles for amusement, it represented something rare in the Japanese academic tradition—a mode of learning whose protagonists pushed forward to new scholarly frontiers without consulting foreign authorities, a discipline which moved from paradigm creation to the formation of an advocate group. Having a vitality all its own, it experienced substantial normal-science growth in relative isolation, largely unmindful of developments in Chinese or Western mathematics.

Western-style mathematics was introduced in the mid-19th century in conjunction with practical matters such as navigation and surveying. The notation of the Western system was completely different from that of the traditional system—a factor that has frequently had paradigmatic significance in dividing one tradition from another. The change from Roman to Arabic numbers and the differences between English-style Newtonian mathematics and the continental, differential notation of Leibnitz had a substantial influence on subsequent development, and the ability of Japanese mathematics to establish its independence from China owes much to its system of algebraic notation. In turn, those who had need of and studied Western mathematics—naval officers and surveyors in the late 19th century, who had been trained at the Nagasaki Kaigun Denshusho (Naval School) and who worked at the Sokuryōshi (Office of Surveying), were very different from the practitioners of *wasan.* Finding it a nuisance to convert all the mathematical notations they encountered in their study of

Western science into *wasan* forms, they determined to use the Western system of notation as was. This choice was made on the basis of convenience, before anyone had asked which of the two systems was superior or thought about blending them.

An examination of the backgrounds of mathematicians listed as teachers in private academies in Tokyo in 1873 reveals the radical split that existed between the two traditions. Almost none of the teachers had had training in both Japanese and Western-style mathematics. After the Ministry of Education opted for the latter in 1872, some *wasanka* (practitioners of traditional mathematics) tried to make the transition, but few were successful—an indication of how difficult it is for a scholar to change his academic style and mode of operation.

Wasan is a part of Japan's cultural heritage of which it can be proud, but as long as its problem-solving methods had to be learned by working on riddle-like problems given by individual teachers, proficiency required a great deal of time. Although this approach stimulated student interest, it could not begin to keep up with the demands of an age which saw the need to train large numbers of scientists and technicians. In this respect it was no match for Western mathematics, where the student attained a solid grasp of basic theory before going on to applied problems. In terms of personnel as well as of notational systems, it was by leaving the tradition of *wasan* behind that Japan was able to embrace Western methods of calculation.

Support Groups for Westernization

Let us now turn to the occupational support groups who successively played a leading role in the introduction of Western science to Japan.

1) *Astronomers and interpreters.* The adoption by the Chinese of the Western "barbarian" astronomy in the 17th century dismayed conservative elements in Japan; however, it promoted in others the desire for a similar change in the Japanese calendar, which was modeled on that of the ancient Chinese. As has been mentioned, the eighth Shogun Yoshimune and his court astronomers clearly recognized the superiority of Western over Chinese astronomy, but these men were primarily government bureaucrats

or technicians whose scope remained limited to their assigned duties of drawing up the official calendar; their interest in Western science was limited to the precision of astronomical data and methods of calculation, and they made no attempt to jeopardize their hereditary posts by entertaining revolutionary paradigms such as those developing in Europe at that time.

Professional interpreters at Nagasaki were well versed in the Dutch language, through which they must have become acquainted with the concepts of Western science. But they also were hereditary government officials who remained within the boundaries of their duty of faithful translation and nothing more. Neither official astronomers nor interpreters published their work for general audiences.

2) *Independent scholars and physicians.* Beginning in the late 18th century, a sizable number of Dutch books containing the term *Natuurkunde* (study of nature) found their way into the country and aroused the interest of various independent scholars who set about translating them, although their foreign-language skills were much inferior to those of the Nagasaki interpreters. The majority of these "Dutch scholars," as they came to be called, were *avant-garde* phsyicians who worked primarily as freelancers, with no strict subordinative links to the governing elite or resultant interest in maintaining an existing status quo. Thus they were not inhibited in stepping out of their line of work and extending their interest to anything Western (except perhaps the Christian doctrine, which was outlawed in Tokugawa Japan). Astronomy was the first area in which scholars sensed Western superiority, but the notion that the West was superior in other fields of scientific endeavor first spread among these independent physicians.

As they inched their way through Dutch texts, they realized that Western science was more than a variant of the natural history line of their own tradition. As they saw it, its essence could be translated as *kyūri* (literally, "investigating the principles of things," a Neo-Confucian term), or "natural philosophy," being a systematic and fundamental investigation and consideration of the nature of things. Thus, although at first there was no clearly established translation for the term "science," *kyūri* was later to become the most common, and, given the vocabulary of the period, it was an informative translation. In recognizing an inquiry into principles at the bottom of such traditional practical studies as medicine and calendar making—in becoming aware, that is, of natural philosophy or physics—they also grasped the hierarchical structure of modern science, from basic to applied. Above and beyond the culture-bound achievements of Western science, they seem to have sensed that it contained a revolutionary paradigm—since the belief in underlying laws of Nature that could be formulated was weak in Japanese traditional thought. This was in marked contrast to the official astronomers who looked upon science as something that could be handled at the technical level and assigned it only a supplemental role.

The physicians recognized the importance of physics and chemistry as the basis of medical work, and several of them founded schools for the teaching of Western medicine. Their students, during their apprenticeship and internship days, moved from one school to another in major medical centers like Nagasaki, Osaka, Kyoto, and Edo (Tokyo), and played a role in diffusing knowledge of Western culture as well as of medicine. While some physicians remained in the practice of medicine, others turned to the teaching of the Dutch language and even to teaching international knowledge and politics. Thus, the physicians' cultural influence was more extensive than that of the astronomers, who limited their activities to the translation of technical works.

But these support groups for Western science were by no means adequate preparation for the reception of the contemporary science. The "Dutch scholars" were still only amateur supporters of uninstitutionalized paradigms, comparable in this respect to members of the Royal Society. Their schools were few and regarded on the whole as *avant-garde*. Had they flourished in the 17th and 18th centuries, their work might have had greater consequences. By the mid-19th century, however, modern science in the West had entered the systematically structured world of the university and reached new levels of normal science advance. Reception of learning that was now systematically pursued in Europe within the structures of the university required the creation of an institutional system around, not self-taught amateurs, but professional scientists who had received a modern education. These early scholars were not aware of the institutionalized aspect of Western science until the mid-19th century, remaining bookish translators of *kyūri*, the paradigmatic aspect of Western science.

3) *The samurai class*. The Opium War in the 1840s between the Chinese and the Western colonialists had created concern among the cognoscenti of the samurai class, but Commodore Perry's visit to Japan in 1853 and the subsequent threat of war unless the country opened to foreigners caused all samurai to realize that the Westerners' science, in the form of their superior technology, was needed for the sake of national defense. As a result, young samurai flocked to the schools of Western learning established by physicians. As samurai were traditionally the ruling class and, in time of war, the warriors, it was natural that their interest in Western science and technology was based on more real and pressing needs than reception of and support for particular scholarly paradigms.

First they tried to learn the Western art of manufacturing firearms, but soon they realized that it was impossible to catch up and compete with Western forces by a crash project of manufacturing cannons. They could purchase hardware, but what was really needed was the software of Western military training and tactics. Moreover, the acquisition of Western military techniques was one of the contributing factors in the overthrow of the ruling Tokugawa family by anti-Tokugawa samurai and the subsequent establishing of the Meiji government in 1868. The samurai recognition of Western science was therefore political rather than cultural.

After the revolution, convinced that Western scientific learning was essential, the ruling class set about disseminating their conviction through the establishment of an educational infrastructure. With government effort and initiative, modern science was fully assimilated in a wholesale introduction through official institutions rather than by piecemeal cultural infiltration, as in the previous era, and modern scientific and technological professions became the artificial creation of the new Western-oriented government.

The main practitioners of the new professions were former samurai. In the past they had received hereditary family stipends in exchange for their loyalty to the feudal powers, the Shogunate or local feudatories. But in the 1870s, efforts were made by the Meiji government to curtail the inherited family stipends of the samurai class as a step towards social modernization. While other classes, farmers, artisans, and merchants, could continue to be engaged in their inherited vocations, samurai completely lost their traditional source of revenue, and had to find new ways of living independently. Since samurai could not compete with other classes in the traditional occupations, like agriculture and medicine, they were attracted to new fields like science and technology. Almost all the early graduates of the engineering colleges were samurai, although the samurai class was only five percent of the total population. Even as late as in 1890, more than eighty percent of Imperial University graduates in engineering and science were samurai.

Thus, Japanese modern science and technology professions were, in the beginning of their formation, very much "samurai-spirited." The samurai as a class were long accustomed by mental habit to think in terms of public affairs and by behavior patterns to playing the game of public office. In contrast with the European pattern, in which science and technology were dominated by the rising bourgeoisie, the new Japanese scientific and technological professions in the last quarter of the 19th century were the province of the proud old samurai class.

The following simple table summarizes what has been said above:

	I	II	III
	Astronomers & Interpreters	Physicians	Samurai
Leading period	18th century	Late 18th-early 19th century	Mid-19th century
Interest	Technical	Cultural	Political
Status	Technicians	Freelancers	Planners & administrators
Role	Judges of superiority	Diffusion & popularization	Institutionalization

The Utilitarian Image of Science

In 18th-century Europe science and technology were distinct

activities with different social origins. In spite of the effort made by the Encyclopedists to liquidate this social interface, the dual structure of science and technology was still maintained, even in the 19th century, by socially separated groups. This was exemplified by such institutional separations as that between *Universitat* and *Technische Hochschule* in Germany. However, there was no particular reason for mid-19th-century Japanese to distinguish between science and technology when facing the impact of modern Western military aggression. To the Japanese it appeared that modern science and modern technology grew out of a single Western tradition. It was not the science-versus-technology dichotomy but rather the tradition-versus-Western dichotomy over which the Japanese were seriously concerned.

While science in 19th-century Europe was still in the main a cultural activity, rather than a practical means of achieving economic growth (as is well illustrated by the issue of the theory of evolution), the Japanese in the late 19th century held perhaps the most modern image of science: it was exclusively utilitarian and pragmatic, planned to enhance the national interest if not purely for profit-making, specialized and compartmentalized. Emphasis was laid on physical and applied science rather than on biology; and hence the style was closest, for that period, to our contemporary scientific technology.

Neither did the debates that were raging in the 1880s in Europe on science versus philosophy and science versus religion have any significant impact in Japan. Since the philosophical and religious dimensions of the Japanese intellectual and academic tradition were attenuated and the sciences were imported in a non-ideological and non-philosophical fashion as established, compartmentalized edifices, the conflict between science and religion never became a major source of controversy in Japan. Western philosophy was considered only one of the newly imported "hundred disciplines." In other words, Japanese modern science was freed from its European philosophical roots: Japanese accepted the paradigms developed in Europe as self-evident and were concerned only with mastering them technically.

Definition of Science

The Meiji encounter with Western culture was in many ways an unprecedented experience. The many academic disciplines of the modern West suddenly burst in upon the country, creating confusion in the intellectual world. A former scholar of Dutch learning, Nishi Amane, sought to provide translations for the array of disciplines and develop a classificatory scheme for all knowledge in his *Hyakugaku Renkan* (Links of a Hundred Sciences), but it was the image of diversity that was to prevail. If Japanese scholars had initially been struck by the fact that Western learning probed the basic principles of things, by the 1860s and 1870s they were marveling at the degree of specialization in which it presented itself. In the Chinese tradition, astronomy and medicine had been recognized as independent professions, but the main current of scholarship—study of the histories and the classics—was the work of Confucian scholars among whom specialization was minimal. Not only did Western learning have an array of fields—chemistry, natural history, physics, etc.—but each had its own group of practitioners.

It was against this experience that the current Japanese term for science, *kagaku* (literally "classified learning"), gained currency. To my knowledge, the first appearance of the word is in the essay "Opinion on the School System," drafted early in 1871 by Inoue Kowashi, who proposed "to invite selected students, teach them the Western languages and then let them specialize in *kagaku*," with the help of foreign professors whom he would employ. In this case, *kagaku* implied not just natural science but any kind of particular discipline, the "hundred specialties," successfully capturing the salient feature of the specialized and institutionally differentiated phenomenon that was a characteristic of science or *Wissenschaft* in the late 19th century.

The suggestion that *kagaku* is a translation of the German *Fachwissenschaft* has also been made. Yet whatever its origins, Japanese of the Meiji period clearly used the term with reference less to the unique methods and paradigms of modern science than to the configuration of differentiated special disciplines in which it manifests itself. Although *kyūri* had represented an understanding of science at the cognitive level, *kagaku* was an attempt to handle it in institutional terms.

In Itō Hirobumi's *Kyōikugi* (Proposals on Education), drafted by Inoue Kowashi and presented to the Emperor in September 1879, we read: "Upper level students should be schooled in the sciences (*kagaku*); they should not be drawn off into discussions of politics." Science was, in other words, saddled with the task of curbing the Movement for Liberty and Popular Rights and training politically docile, narrowly specialized professionals. Thus the very same "science" that, as *kyūri,* had been considered a philosophic paradigm during the Edo period (up until the mid-19th century) was perceived by Meiji eyes as an array of non-philosophical, technical disciplines.

Emphasis on Physical Science and Specialization

Japanese pioneers in Western science during the late Tokugawa period were impressed by the Western process of inquiry into natural philosophy. They came to feel keenly that, although there was no great gap between East and West as far as the classificatory knowledge of natural history was concerned, Chinese and Japanese culture lacked belief in the underlying regularity in Nature that underlay the "investigation of its principles," namely natural philosophy. This *kyūri* aspect was now, under the Meiji *kagaku* scheme, narrowed down to the physical sciences and scientific technology. The primary school curriculum, therefore, as prescribed for the first time by the Ministry of Education, was not natural history- and biology-oriented, like that of American primary education, but physics-oriented.

During the 1870s and 1880s greater emphasis was placed on science and technology in the Japanese educational curriculum, from elementary school to university level, than in that of any other nation. For instance, mathematics and science occupied about one-third of the school curriculum at the lower grades (first four years) and two-thirds at the upper grades of the eight-year elementary education—though due to the shortage of qualified teachers it is somewhat questionable to what extent these ideal plans were put into practice. At the university level too, the emphasis on science and technology was evident in the high percentage of graduates of Tokyo University in scientific disciplines (85 percent in the 1880s as compared to 40 percent in the 1920s).

Official and Planned Character

At the frontiers of newly forming disciplines in 19th-century Europe, the scientist was free to take up for research any problem that interested him, and scientific communities were formed by individual scientists drawn together by a common interest. The voluntary activities of scientific or professional societies usually preceded their inclusion in the university curriculum. In the case of Meiji Japan, however, and as is generally the case when a foreign discipline is artificially transplanted under state sponsorship, this process was reversed. The government first created institutions for training specific personnel, and only then did university graduates in each discipline form their scientific societies, not purely for academic purposes, but mainly with common interests in their new and still very weak scientific careers. Created in this way, the scientific community in Japan had a "planned character"—planned for the specific purpose of catching up with the Western standard of science as quickly as possible.

In the 19th century, established precedents or formulas for a national science policy were difficult to find even in the advanced nations. Thus, the Meiji government had to find its own way by trial and error. Unlike today, it was simply unimaginable to have some sort of international aid or technology transfer from a developed country to a developing country. In Europe and America science and technology were not yet state-supported but largely activities within the private sector. The Japanese government was simply trying to purchase and procure the scientific facilities and manpower available in the European and American free markets.

On the practical level, first of all, for building a modern state qualified teachers and engineers were urgently needed. Here the new government, in pursuit of its aim, built schools and factories, trained scientists and engineers in short and intensive courses, and sent them off to their posts. Rather than having every scientist following his own research interest, as was often the case in Europe and America, priority was given in a collective way to certain basic tasks, the accomplishment of which was indispensable for the operation of a modern state: geographical and geological surveys, standardization of weights and measures, meteorological observations, sanitation, printing, telegraph and telephone, military works, railways, surveys of natural resources, etc. All these activities were carried out by the Ministry of Technology, the Ministry of Interior, the Ministry of Finance, the Commission for the Colonization of Hokkaido, the Army, the Navy, and other functioning governmental agencies under the supervision of many foreign engineers working in Japan. To conduct such ambitious nation-wide projects, these agencies had to have their own short-course training programs to provide field assistants for imported advisers. An example was the Telegraph School of the Ministry of Technology.

Besides this, the government entered into private entrepreneurship, constructed and managed pilot plants, and guided and subsidized new kinds of industries. The Ministry of Technology and the Commission for the Colonization of Hokkaido were the two major institutional innovators; they carried out new experimental programs and were also the centers of Westernization and modernization.

These "public science" enterprises were from the beginning exposed to financial risk. Many of the projects of the Ministry of Technology eventually proved to be too far ahead of their times, as they were intended to introduce the technology of an industrialized society into a pre-industrial environment. For instance, railway construction was economically unsuccessful at first; it only paid off commercially after 1880, in the second phase of industrialization. Fukuzawa Yukichi, Japan's foremost exponent of industrial revolution, concluded: "We should not blame them too much for their financial failure. After all, it was a costly tuition for the Japanese to learn civilization."

In giving priority to the construction of an institutional system within which to transplant Western paradigms, Meiji Japan paid more attention to the configuration and format of learning than to its content. Scholars troubled themselves little over how new scholarly paradigms were being born. Neither did they entertain the notion of participating in the ongoing advance of normal science. Their first preoccupation was the creation of an institutional framework to house knowledge previously canonized and accepted as standard in other traditions.

The Priority of Institutions

The introduction of Western learning as a system left little room for a comparative examination of paradigms in particular fields. If the learning of the West was advanced, this was entirely due to the fact that the West had undertaken to pursue scholarship in an open, regularly organized, institutional way. The first imperative then was the adoption of the Western type of scientific institution. This conviction and commitment underlay the introduction of the sciences *en bloc* and inspired wholesale Westernization in the 1870s and 1880s.

Figure 6–1 displays graphically the contrast between paradigm-initiated science and institution-initiated science. The left side depicts the Meiji Japanese case in which institutions were constructed first. In contrast to the gradual spread of scholarly traditions from one country to another by voluntaristic support groups, government policy aims are dominant in this second pattern. Scholarship, in a word, comes under state management. One might say that Japan was the first nation to experience the worldwide trend toward state-managed science with which we are familiar today.

Institution-building seemed like an obvious goal to the first generation of university men, and they poured themselves into the task. By the late 1870s, universities were being built with desig-

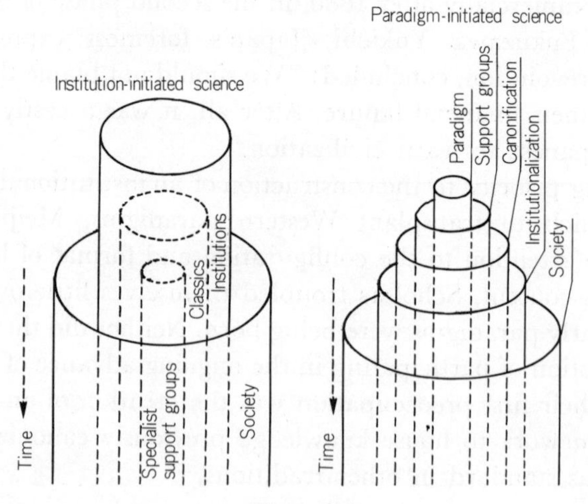

Figure 6–1

nated quarters for the specialized disciplines of the West, and specified numbers of students were allotted to each.

The "Dutch learning" scholars of the Edo period could feel that they were at the center of cultural contact and interflow, and pivotally involved in the vital task of enlightening their countrymen. In this spirit they held "mini-expositions" to which each brought knowledge and information to share with others. While this Royal Society-type atmosphere could still be found during the first Meiji decade of the 1870s, scientists born during the new era were a different breed. Bearing the hopes of the nation on their shoulders, they lived in assigned rooms where nary a truant glance was permitted to interrupt their single-minded devotion to the transplantation of Western scholarship. For this task the fine divisions of the professorial "chair system" proved both opportune and efficient.

Under these circumstances, the academic societies that developed in Japan were certain to present some distinctive features—features that allow us to contrast them with those formed on scholarly frontiers. Academic groups in the "Dutch scholar" tradition embraced Western learning in general, but their contributions were limited to persuading their countrymen of Western cultural superiority, providing Japanese terminology for Western concepts and methods, and offering initial guidance in the classification of knowledge. They had neither the capacity nor the background to turn themselves into a support group for particular scholarly paradigms. By 1877 their role in awakening their countrymen to the riches of Western civilization was finished.

In the decade which followed, these men founded the Tokyo Gakushikai in a conscious attempt to emulate the Royal Society. This body was never to acquire the kind of authority that had traditionally belonged to such central academies in the West and which was, at that very moment, under attack from specialized academic associations in Europe. Several specialized academic societies also appeared in Japan at this time, but in a nascent academic world oriented toward the importation of knowledge from abroad, these new societies were typically built around a nucleus of scholars who had studied in the West, and were composed of scientists trained in Japan by foreign teachers. Ascertain which came first, academic associations or academic departments, and you will be able to determine whether a tradition is of the normative type, in which scholarly groups are typically formed on a voluntary basis, or the countertype in which institutional structures take precedence. Because in Meiji Japan the creation of institutional structures preceded the formation of scholarly groups, these societies acquired a guild-like character and an alumni-association atmosphere from the very beginning.

In drawing from the well of Western knowledge, it seemed most efficient to attach a few small hoses to the Euro-American pump and channel specific bodies and forms of knowledge directly, in as pure a state as possible, into the several university department "fish tanks." One big hose might have drawn in a spectrum of ideas so wide as to have produced hesitation and vacillation, while too many hoses would have resulted in confusion. Energy and budget could not have been used very effectively in either case. Scholarly paradigms normally originate in open markets. In fish tanks fed by narrow hoses only normal-science development can take place.

Once the institutional framework is set up, the textbooks written, and the scholarly community in possession of the means to reproduce itself, then research can begin. As long as the system continues to operate smoothly, normal science progress is assured. At the very least, the knowledge and methods that enter will continue to be passed on. Should the well run dry, however, the flow of normal science problems in the hose will slow to a trickle. Eventually the hose will stiffen and the fish tank it feeds will empty—at least of any intellectual nourishment. There are only two conceivable remedies for this situation. If a fresh spring can be found, a new hose can be installed. This is, however, not as easy as it may seem, for the old hose is part of a whole established network of hoses that cannot be tampered with at will, even if only to replace one of them. Alternatively, scholars might consider providing their own nourishment and attempt to become involved in the making of paradigms themselves. To do this, however, they must abandon their hose-fed fish tanks and plunge into the great wellspring of human learning—something that scholars artificially raised within specialized institutional compartments are ill-equipped to do. Turned loose without a familiar paradigm to guide them, they are quite lost.

Yet this great wellspring remains the natural habitat of scholars. Here they are called to gather in voluntary groups around the paradigms that emerge from their midst, and to disperse in search of new ones when the old lose their capacity to generate fresh problems. When the proponents of the scientific revolution realized that the rigid structures of their tradition were an obstacle to further advance, they tore them down and built new ones better suited to the new paradigms they were creating. Circumstances at the present time are quite different than they were in the days when the Royal Society was founded. Contemporary institutions serve to ensure the continued advance of the kind of normal science to which the insights of the 17th century gave rise. Faced with a new scientific revolution, however, these same institutions can become structures of oppression, for though by nature free, scholars have come to enjoy the protection of the state and allowed themselves to be ensnared by the system.

The Receiving Nucleus

Reception of a paradigm from an alien culture always takes place around some kind of nucleus that serves as the initial and initiating agent or factor in the receiving process: a mode of thought, a scholarly group, a set of institutions, a basic policy orientation, or the structures and policies of a colonial power. The degree to which and manner in which a paradigm is accepted from abroad vary according to the receptivity or resistance of this nucleus and its particular proclivities. In other words, the nature of the nucleus gives rise to differences between the original paradigm and the structure of learning associated with it after it has been received and reformulated.

For instance, the Islamic world developed a tradition that combined Indian and Chinese influences with the Greek tradition. There must have been something in the world of Islamic thought and theology that facilitated the assimilation of these foreign influences and traditions. Yet a closer examination reveals that not all intellectual groups were equally receptive. As a group, theologians actually opposed the influx of crucial writings like those of Aristotle. It was rather the philosophers and the physicians who undertook to translate foreign works and who were chiefly responsible for the importation of foreign culture.

Meiji Japan presents a different pattern and problem. Much ink has been spilled over the question of whether Japan's modernization was the result of external pressures or whether it is to be traced to structural changes in pre-Meiji Japanese society. Insofar as modern science is concerned, the answer seems clear: the intellectual life of the Edo period could not have generated modern science of itself. Autogenesis might have been possible if there had been more time, though the result would almost certainly have been a modern science quite unlike that developed in the West. The rapidity with which Japan adopted modern science is best understood not in terms of internal "push" factors that might indicate Japan was moving in the same direction as the West, but by the existence of a peculiarly receptive nucleus.

Nuclei, Receptivity, and Reception

Nagasaki interpreters played an early and transient role in the introduction of Dutch learning during the Edo period, but the nucleus of Japan's initial encounter with modern science was eventually drawn from two traditional professions: the official astronomers (*tenmonkata*) of the Tokugawa government and the relatively independent physicians. The *tenmonkata* tended to confine themselves to satisfying official demands for such Western astronomical knowledge as might be useful in making calendrical revisions. In the physicians one finds a group of voluntary supporters who were free to develop their interests beyond immediate medical purposes and press on to general examination of the character of Western learning. As foreign pressure mounted after 1840, Shogunate officials showed an interest not only in Western military technology but in the political institutions and social organization of the West as well.

Meanwhile, in China, direct encounter with the West led to the bitter experience of the Opium Wars and contributed to the emergence of Tseng Kuo-fan, Tso Tsung-tang, Li Hung-chang, and other Westernizers who became the most active element in the Chinese nucleus. The introduction of Western learning was initially part of a program to strengthen China's maritime defenses; for this mathematics and its applications to engineering were particularly necessary. Western superiority in this area had been recognized ever since the arrival of the Jesuits, but there was a need for modern instruments. The Chinese also needed practical military technology and the physics and chemistry on which it was based. To supply this knowledge, Westernizers at Shanghai's Kiangnan Arsenal and elsewhere undertook extensive translation work. Advocate groups for Western medicine and the dissemination of modern science as a mode of thought were, however, slow to form. The Protestant missionaries were to have the field of Western medicine almost to themselves throughout the 19th century.

But whether in Japan, China, or even a colonial territory such as India, there seems to have been an important difference in the reception of modern science and technology between nations that had nuclei within their own cultures and most other regions of Asia and Africa. This contrast can be seen clearly in the former Japanese colonies of Korea and Taiwan. In Korea, reception of the wave of modernization was part of an intense desire for national and cultural independence, while in Taiwan the emphasis has been on the materialistic dimension of modern scientific and technological civilization.

Intellectuals have typically been in the vanguard of nationalist movements and, at the same time, supporters and advocates of modern science and technology. Wherever their attempts to introduce modern science have met with resistance rooted in traditional paradigms, the ensuing conflict has had a major effect on the way in which that science has been eventually understood and received. In some places traditional paradigms impeded the acceptance of modern science. Yet in China this kind of conflict proved to be the setting for the emergence of a modern science that was more than an assemblage of techniques for improving the material aspects of life, a science whose paradigms were to become the ideological rivals and destroyers of traditional views of the world. The notion of China as the unique center of civilization did resist the acceptance of modern science, but science would eventually do battle with the Confucian tradition, even providing ideological legitimation for the smashing of Confucian temples. Where traditional forms of nationalism have failed to encompass modern science, they have not become modern nationalisms.

Where colonial policy was the central factor (for example, in the former Dutch territory of Indonesia), the dispatch of missionary doctors to serve white settlers was the first step in the coming of modern science. Following this, two other problems spurred its introduction: the need to train the local citizenry in basic hygienic techniques in order to prevent epidemics among plantation workers, and the need for lower-level civil servants in the colonial administration. Toward the end of the 19th century, as the number of enterprises capitalized abroad increased and colonial policy came to be concerned with the investment of surplus capital in underdeveloped areas, administrators began to think in terms of giving the local populace training in agricultural and some engineering techniques. In carrying out this policy, it was in the rulers' interest to excise the ideological elements historically associated with science and technology (e.g. the assumptions of the French Enlightenment), to avoid fields that tended to spawn anti-establishment ideas, and to give the locals training in purely technical fields such as medical care. This type of education also served local needs, at least in the sense that the native population had more use for immediately applicable manual techniques than alien studies based on the hegemony of Western culture such as European literature. Where there was no nationalism, modern science and technology tended to be diminished and exploited to the mutual benefit of rulers and ruled.

The State Selects

Meiji Japan responded to discovery of the West's academic system with the creation of its own. As we have noted, this attempt to duplicate the disciplines of the West as a set was not geared to scrutinizing particular modes of learning and making substantative choices among them. Still, Japan did manage to maintain its independence and, with the closed academic market that had been part of the special relationship with the Dutch a thing of the past, was in a position to make certain choices from among the European Western models. In a short time, the government had shown a distinct preference for English technology, American agriculture, and German medicine and natural science.

In the Meiji government's policy of Westernization, the establishment of Western-style educational institutions occupied a central position. The samurai leaders surveyed the Western system country by country, and in 1870 drafted principles for the dispatch of students for study abroad. The following subjects and countries were listed:

Britain:	Machinery, geology and mining, steel making, architecture, shipbuilding, cattle raising, commerce, poor-relief.
France:	Zoology and botany, astronomy, mathematics, physics, chemistry, architecture, law, international relations, promotion of public welfare.
Germany:	Physics, astronomy, geology and mineralogy, chemistry, zoology and botany, medicine, pharmacology, educational system, political science, economics.
Holland:	Irrigation, architecture, shipbuilding, political science, economics, poor-relief.
U.S.A.	Industrial laws, agriculture, cattle raising, mining, communications, commercial law.

Looking back on the history of 19th-century science, the above assessment of the strengths of the Western countries seems, by and large, correct and objective.

As for the educational system as a whole, traditional ties with Holland meant that Dutch institutions were studied first, but as the Education Ordinance (*Gakusei*) of 1872 was being readied, the education ministry undertook a wide-ranging survey of European and American education. The upshot of this study was that the early Meiji educational system borrowed the centralized, pyramidal structure of the French system in externals while being largely American in curriculum and texts. After the "political change of 1881," however, the "tilt" toward Germany intensified. New inquiries instigated by Mori Arinori and Inoue Kowashi in preparation for the creation of the Imperial University led to a heightened appreciation of German scholarship, and a consensus began to build in government circles in favor of the German tradition. Japanese scientists studying in Europe were also increasingly sensitive to the reputation of German science, and cases in which scholars shifted their studies to Germany from England (or elsewhere) were not unknown.

The fact that the Japanese university system achieved its definitive form during a period in which German higher education was being imitated in many parts of the world meant that the assimilation of the American graduate-school model would be deferred until after World War II. The Harris School of the Natural Sciences, set up at Doshisha in Kyoto in 1889, was modeled on the American system of graduate education, but it failed to attract a sufficient number of students and was discontinued in 1896.

To this point my discussion of the acceptance of modern science has been chiefly concerned with the importation of existing knowledge. Since the act of acquiring Western knowledge is not of the same order as producing something new, it cannot be called scholarly or scientific activity in the strict sense of those terms—the tremendous amount of effort that went into mastering foreign techniques and methods notwithstanding. But where does independent research begin? How does learning in the complete sense of that term—studying, questioning what has been learned, and bringing forth new knowledge—begin in non-Western lands where modern science must be initiated by the transplantation of foreign paradigms?

Local Science

Wherever modern science has been accepted not simply as positive knowledge but as a mode of scholarly practice, research has begun with "local science." By "local science" I mean, in the first instance, studies that focus on problems with a particular relationship to or meaning for a specific country or region as opposed to the more general and universalistic problems that typically occur in fields like physics and chemistry. In such activities as the classification of native plant and animal life and the study of endemic diseases, "local science" was being practiced by local inhabitants prior to the advent of Western scholarship, although the task of reordering these fields in accordance with the paradigms of modern science was initially the work of European and American scholars. The process by which the practice of modern scholarship came to inhabit and influence the non-Western areas of the world can be divided into three stages: 1) the collection of materials for introduction to the Western academic community, 2) survey projects related to local development, and 3) education of the local population.

1) *The gathering of materials.* Colonial expansion offered European intellectuals an opportunity to satisfy their curiosity. Scientists learned about the explorations and discoveries of warships and commercial vessels from medical officers and crew members, and were eventually allowed to go along on these voyages themselves. Westerners who came to undeveloped regions as colonial bureaucrats, merchants, doctors, or missionaries also contributed to the accumulation of knowledge about these areas. Later they were joined by others whose interests were more purely scholarly. Naturalists seeking species of flora and fauna unknown to the Western intellectual community extended their field to overseas areas, as Charles Darwin did in his journeys on the Admiralty survey ships *H.M.S. Beagle* and *H.M.S. Challenger*. In the United States, scholars joined the wave of frontier development and westward migration, and then expanded their horizons from the Pacific Coast to Hawaii and on to East Asia. Meanwhile Japan, which had become known in the West through the writings of men like Kaempfer and von Siebold as the "closed country," began to attract the attention of many scholars.

With the competition for colonies, the writings of Orientalists proliferated, and collections of material about the racial characteristics, languages, customs, and morals of local populations became a recognized genre. Unlike physics, which seeks after general truths, this was a local science, more like the gathering of data about local flora and fauna. Those who engaged in it reported their findings to the Western academic community and published numerous journals dealing with things Oriental, but they did not put down roots in the regions of which they wrote or seek to plant a scholarly tradition there. Where advocate-support groups did happen to emerge (e.g. in Japan where a group of "Dutch scholars" gathered around von Siebold), one can see a beginning of Western-style scholarship.

2) *Survey projects.* As we have noted, the 19th century eventually led the European powers beyond simple mercantilistic plundering to a concern with productivity and the development of

full-blown colonial policies. This change in attitude gave rise to the need for basic surveys of the potential of colonial lands—studies that provided a new outlet for "local science." One of the pioneers in this activity was the Englishman Thomas Raffles (1781–1826). As a colonial administrator in the East Indies and Singapore, Raffles' interest in local conditions went beyond the Orientalist's fascination with the exotic. Gathering information on natural resources, mining products, and local flora and fauna with an eye to formulating colonial policy, he initiated a type of inquiry that was more carefully done and more systematically focused than earlier writings.

A global observation network was established in astronomy, including an observatory in Capetown, South Africa, for observing the heavens in the Southern Hemisphere. It inspired a number of mid-19th-century projects aimed at a world-wide network for geophysical observations. Local studies of endemic diseases were also seriously begun. The Dutch East India Company had employed doctors for white residents and settlers, but the ineffectiveness of Western medicine against some local diseases prompted improvements in hygiene and led to the establishment of a research institute for contagious diseases that was one of the first to adopt the new, bacteriological approach to the study of these maladies.

3) *Educating the local inhabitants.* If this kind of organized research was to be successful, it required the assistance of native personnel. For this and other reasons, it became necessary to educate at least a small number of local people. In other cases, however, Western education began at local initiative. In either event, once local inhabitants became familiar with the goals and techniques of modern scholarship, scholarly activity could no longer be initiated exclusively by Westerners. "Local science," begun as a response to the impact of Western scholarship, took on a life of its own.

As we have seen, response to the Western stimulus was varied. Japan managed to stave off for an extended period the activities described above as belonging to the first stage, and after the Meiji Restoration kept the second and third stages under Japanese control. Where there was full-fledged rule by a colonial power as in Indonesia, the Philippines, and Japan's former territories, the ruling power was in charge at all three stages. In intermediate situations, such as those of China and India, where an ancient cultural heritage provided fertile ground for strong nationalism, local interests strove to set up their own modern educational programs and engage the native scientists trained therein in the survey work of the second stage.

If the ideal situation for modernization is one in which advanced countries exert a scientific and technological influence on underdeveloped nations but do not engage in acts of colonial domination, then academic independence in these states is best served when advanced countries promptly pull out once they have started underdeveloped nations on their way to modernization, have trained a local support group for modern science, and have supported institutions that make it possible for such a group to replenish itself.

Local science, then, is an offshoot of the scholarly paradigms of the West and is built upon them. It is normal science development made possible by applying proven methods, approaches, and questions in a new social environment. In these local science (or field science) activities—the discovery of new species of plants and animals, the making of latitudinal and longitudinal surveys, the construction of geological charts and maps, the study of endemic diseases, the analysis of native products—local scholars have many advantages. Almost all the studies of Japanese scholars that proved to be of interest to international academic societies during the Meiji period were of this type.

Local science is often directly related to the national interest. It includes research that international custom expects a modern state to provide. Can one imagine a modern state without latitude and longitude determinations and a geologic map? For this reason local science often receives heavy institutional support from establishment sources. While the Nazi regime in 1930s Germany was rejecting internationalist "universal science," calling it "Jewish science" or the "white international" and driving scholars like Albert Einstein into exile, at the same time, scholars engaged in local science managed to link their work with national interest and Nazi ideology, and went about their work looking quite magisterial.

Since local science involves the application of an acquired

method to local subjects, once a researcher establishes himself in his field, he can turn out a steady stream of normal science work. In non-Western countries, the field sciences are the first to overcome a sense of inferiority toward the West and to produce meaningful work of their own. Sometimes they even yield local paradigms. In China the field science of geology was the first active modern science discipline, and one of the first to develop its own academic association (a Chinese geological society was formed in 1922).

Can achievements in a local science have international significance? If "international significance" is defined as being cited in the work of foreign scholars and affecting work done in foreign countries, then Japanese work in studies of the Japanese physical and social environment can be so considered. Since such local research influences syntheses and generalizations formed outside the Japan field, it contributes to general paradigms for the study of nature. But the limits are fixed: insofar as local studies result from the application of Western paradigms to the Japan "field" they cannot transcend normal-science research.

Universal Science

Local paradigms have local support groups, while international paradigms have international groups of supporters. One can measure this dimension of any given scholar's work by the geographical distribution of those who publish his work or cite him. In the perfectly international case, a scholar whose work was known, for instance, to five Japanese would also be known to ten Americans, ten scholars in the Soviet Union, and ten Europeans. In other words, his reputation would reflect the international distribution of the limited number of scholars in his specialty. Thus we would expect him to be, as the saying goes, "more famous abroad than in Japan."

In universal disciplines such as mathematics, physics, and chemistry, which have no local fields and are defined by method rather than subject matter, paradigm-importing countries initially find even normal-science achievements no easy matter. Scholars in these nations may dream of the day when they will have arrived by unstinting effort at the front lines of scholarship and are ready to engage in original research, but scholarship will not stand still while they try to catch up. There have been many instances when a Japanese researcher thought he had done something new, only to find that it had already been done in Europe, particularly in an earlier age when the round trip between Japan and European centers of science took several months by ship. Under such geographically isolated conditions research naturally gravitated to whatever problems happened to have been overlooked by Western scholars.

But the intellectual environment itself is an even greater problem. Scholarly activity can be sustained only *among* people. One can readily grasp where the frontiers and what the problems and issues are through contact with people at the forefront of one's discipline. But when a person can find no one to talk to, his ideas do not easily mature. Perhaps this explains why almost all the international-level work done by Japanese scientists during the Meiji period took place while they were abroad: after returning to Japan many of them seem to have retreated from the front lines of scholarship.

Japan's second-generation modern scientists were relieved of the burden of institution-building and began to concentrate their energies on scholarly research. As they embarked on this undertaking, however, they were made painfully aware of the distance between their own work and what was going on in the West. Lacking the self-confidence to venture upon such an "awesome" task as paradigm-building, they began to do little "studies," picking up whatever grain was left after the harvest of Western scholarship. Following paradigms created in Europe, they pushed forward with normal-science research. This was second-grade work, but it was part of the scholarly mainstream, and not without meaning. Moreover, as rewards for the practice of normal science improved, research along standard international guidelines may have become the path of least resistance. For one thing, since the paradigm is already set out, methodological confusion is eliminated. Then too, it is easier to produce reputable work if one is confident about participating in the mainstream of world science.

When J. D. Bernal spoke of Japanese science (as he did in *The Social Function of Science*) as pedantic and lacking in imagination, he was commenting on a tradition that had found its institutional patterns ready-made and appeared to be dominated by multitudes of small fish protectively nurtured in its institutional fish tanks. He was voicing his inability to find, amidst the pressing crowd of little fish, a significant number of individuals big enough to tackle the work of paradigm-building.

As the case of Terada Torahiko reveals, however, this cozy environment may be uncomfortable for men with their own ideas. Terada was not interested in joining his physicist colleagues in following like a school of minnows after European scholars. Persuaded that "so long as Japanese physics continued to look at nature solely through the eyes of Western scholars it would never advance," he chose instead to undertake local geophysical studies of earthquakes and develop a group of research activities on local problems collectively known as "Terada physics."

The reason Japan seems to be an academically backward country is that its many professional scholars have devoted themselves to resolving normal-science problems without a significant effort to formulate new questions. Scientists confident they are operating at the frontiers of their discipline are forever looking ahead—scanning the horizon and thinking about where to go next. They are always prepared to change their course and to being their support group along with them. Those who follow behind are unwilling to think about reorienting their work until they have reached the forefront of their fields. They fear that giving any hint of moving in a new direction would be like admitting failure. It was perhaps this mind-set that prompted Terada's fellow scientists to scoff and sneer at his "physics." My personal impression is that many Japanese scholars, even those in the prime of their academic lives, speak earnestly of "applying themselves conscientiously to their studies" and maintain a self-consciously pious attitude toward their work, while in places where researchers are constantly striving to break new ground, even the graduate students seem full of a venturesome spirit which says, "Let's tackle that question: it looks interesting!"

Young scholars in Japan who have yet to acquire a mature identity do not venture into unknown territory where the conditions for creativity are not apparent. They put their hearts and souls into solving problems posed in advanced countries. They are quite literally a "support" group; they do not put forward paradigms. Thus the history of modern Japanese scholarship lends some credence to Hitler's suggestion that Japan's scientific and technological civilization could be described as "culture-transmitting," but not "culture-creating" (*eine "kulturtragende" aber niemals als eine "kulturschöpferische"* [*Mein Kampf*, p. 319]). This passage was omitted from pre-war Japanese translations). But is it impossible for academically backward countries to formulate an original agenda?

The in-depth knowledge that modern normal science yields can obviously not be attained without effort. Yet if it does nothing else, the history of science and learning should disabuse us of the notion that the development of modern science and, more broadly, the growth of the Western academic tradition, have been direct and logical progressions such that each step was dependent on having successfully negotiated the previous one. There have been many scientific revolutions, and each has brought a change in the goals and direction of scholarship. In the wake of these changes, the obsolete normal science has often seemed like a great waste of effort. Is it impossible then that scholars in academically backward nations, who are as yet not overly burdened with the institutional traditions and professional groups of normal science, should forgo slavish devotion to the extension of conventional normal science and set themselves to the task of scholarly renewal? The most likely places for future paradigm-building are precisely those quarters which have abandoned the effort to compete with the "big-science" money and organization that has dominated advanced countries in recent decades. Paradigm-building does not necessarily require money.

As Japan's second generation of internationally oriented scientists appeared on the scene, government patronage and aid dropped off sharply. Construct a giant telescope that might contribute to mankind's understanding of the universe? Let countries that have a special interest in such things build one; it won't be done in Japan. We cannot afford to supply our scholars with such luxuries. In terms of national interest, local science is quite sufficient. As far as the State is concerned, the production of scholarship at international levels is significant only for appearances' sake.

Even appearances became less important as the unequal treaties with the West were successfully revised and the urgency of appearing "civilized" to impress Europeans lessened. Involvement gave way to a policy of *laissez-faire* under which government regarded science as something in which it should not become involved directly. A delicately balanced academic freedom came into being, and "academic science" put down roots. By the 1910s it was professional scientists who, in an attempt to stress the significance of their own work, were advancing the argument that the level of Japanese scholarship was a matter of national "face." Because of the dignity associated with the word *daigaku* (university), university-related research institutes preserved their singularly high status long after their counterparts in the United States had been equalled by government and business establishments. Though the mobilizations of World War I had already called "industrialized science" into being in Europe and America, Japanese government and business circles remained slow to recognize the possibilities that lay in this direction. A minority of technocrats oriented toward "industrialized science" did emerge in the 1920s and 1930s, but the war was over before they were able to demonstrate their capabilities.

What is to be done? How does one make tankfuls of malnourished small fish big and strong? One way is to turn them loose in the ocean, that is, to give them an opportunity to participate in the international academic world. Another is to catch big "international fish" and put them in the tanks with the small ones. Although the scholarly growth of individuals like Takamine Jōkichi and Noguchi Hideyo suggests the positive effects of extended periods of research abroad, this approach has, in general, not contributed to the growth of the Japanese academic tradition or to the improvement of the domestic academic environment. Those educated in other countries have more often had little influence on their institutions after they returned to Japan. Let us then consider the second alternative.

In the previous chapter, I attributed the initial success of American science to the large volume of "B"- and "C"-grade scientists it produced. This feat required nothing more than a stable institutional setting. In other words, the United States, like Japan, started as an academically backward country and began by building institutions. As elsewhere, further advance would require that the seeds of new paradigms be thrown among these "B"- and "C"-grade scientists. In America, the influx of first-class scholars from Europe and their participation in indigenous research teams played a vital role at this stage. The United States benefited greatly from the peculiar conditions that prevailed in Europe after the rise of Fascism and the outbreak of war; but regardless of the circumstances that brought them there, the combination of top European theoreticians with the best facilities money could buy, and plenty of manpower enabled America to build the atomic bomb and fashion artificial satellites. Whatever one may think of these products, the important point is that by adding first-class Europeans to their own research teams and continuing to expand public support of science, Americans finally liberated themselves from their long-standing "European complex," and now produce most of the world's top-flight scientists.

Before the outbreak of World War II, there seems to have been a plan afoot at Tohoku Imperial University to engage Albert Einstein as a university professor. Although this plan never materialized, Einstein's coming might well have made a big difference in Japan's academic culture. Through intimate contact with one of the world's leading scholars would we not have understood what it was to be the best and, at the same time, gained enough confidence in ourselves to believe that if this was the best, we too could aspire to it? Surely it would have been a great challenge and—more importantly—of great psychological significance in liberating us from our sense of inferiority.

Western Studies

We have examined the development of local science in non-Western countries as an extension of Western paradigms and the struggle of non-Western scholars to participate in the universal sciences. As a third case let us now consider local studies as they are pursued in Japan. Omitting, for the present, studies of Asian and African regions where modern academic groups are as yet not fully developed and Japanese are cast as "advanced-nation scholars," we shall focus on Japanese studies of the West.

Students of the West in the mid-19th century were, in a certain sense, in a happier position than the later generations of Occidentalists for whom scholarship would mean modern academism. The earliest scholars, sustained by awareness of their pioneering role as leaders and teachers of a people not blessed by science and civilization and ignorant of what was going on in the world, wrote solely for their fellow Japanese. Having little interest in what anyone else might think, and nonconformist by nature, they were not burdened by that sense of inferiority toward Westerners that is so much in evidence among later academics.

Yet as scholarship became an institutional affair dominated by formalized instruction and transmitted by persons known as "university professors," Japanese scholars began to operate within the framework of an international "merit system." Of course the Japanese university was far from being a free marketplace open to the world, inundated by foreign scholars competing for Japanese jobs. The walls of the fish tanks were well built and impermeable to outsiders. Still, the achievements of the larger world were accessible—not just to a few individuals but more generally. As Japanese scholars became conscious of international standards, they could not strut around like cell-block bosses even when their local authority went unchallenged.

Natsume Sōseki's London diary gives us a graphic picture of how this predicament tormented him. A university graduate in English literature with a secure future as a professor ahead of him, he felt that he could not hold a candle to England's own specialists, and tried to comfort himself with the thought that at least he knew more about English literature than the lady who ran his boarding-house.

This was a new anxiety, one that does not seem to have seriously disturbed either the generations of "Chinese scholars" who had been engaged in the introduction of Chinese culture since Nara–Heian times, or the "Dutch Western learning" scholars of the Bakumatsu period. With Sōseki's generation, the notion that the scholar is obliged to know what international standards are and to produce work at that level began to impress itself strongly on the Japanese academic mind.

Acknowledgement of international standards implied at least tacit acceptance of the values of modern scholarship, including the notion that the significance of any given piece of research is to be judged in terms of its contribution to the world's storehouse of knowledge. More especially, it meant that the Japanese academic world was on its way to participation in an intellectual environment that began when normal science became the model for a broad spectrum of disciplines in the 19th century—an environment in which scholarship is much less a matter of persuading, teaching, or making a name for oneself among educated people generally than it is of writing scholarly papers and fighting to be first with a new discovery. How long this environment will exist and what standards will be like in the future are open questions, but this is the understanding of academic inquiry that has sustained modern science and modern scholarship for more than a century.

Geology has a local dimension. Spheres of influence remain clearly marked off in this field, and Japan is most unlikely to dispatch a survey team to study the geology of England. Yet inaccessibility of the European continent is not a major impediment to the study of geology in Japan. Lyell's paradigmatic *Principles of Geology* may have drawn heavily on the geology of England, but the modern discipline to which it gave rise was also a natural science committed to the search for general laws. Thus Japanese scholars do not need to make a special effort to conduct studies in the already well-trammeled field of European geology. They can develop the relatively less studied geology of Japan and the Asian region and seek general laws here just as well. Moreover, inasmuch as it is nature that is being studied, there can be no question of one inquiry being "culturally superior" to the other. Thus a scholar in this field has no reason to feel inferior even though he might not be well versed in the geology of Europe. Archaeology is similar in this respect.

But in the case of Western history, Western philosophy, or European and American literature, things have not been so simple. These subjects were initially incorporated in the early Meiji curriculum not simply because they were being pursued in Western universities, but in order to provide an institutional outlet for the dissemination of the kind of basic knowledge of and tools for learning about the West that had been a part of the "Dutch learning" tradition.

Yet when the early goals—compilation of dictionaries and translation of most of the scholarly classics—were attained, what were the teachers, translators, and interpreters of Western culture to do? Could they make the transition to scholarly research in the modern sense? A variety of factors handicapped their efforts to do so, the chief among them being the language barrier, the lack of documents and materials, the lack of a community of scholars with similar interests, and a sense of the remoteness of the subject.

The most striking feature of modern scholarship is the premium it places on the production of scholarly books and articles. If I were an official of a global foundation for scholarly research, I would be unlikely to fund a Japanese scholar who proposed to do research in medieval Latin literature. Instead, the grant would go to a European who had been studying Latin since childhood and would simply be able to do research in this area more efficiently. This way, everyone would agree, the foundation would be getting more for its money.

Specialists in Western studies are no doubt painfully aware of this difference. (Any who remain unfazed by the weight of indigenous scholarship shall not concern us here: they lack the sensitivity one expects in a scholar.) Yet exposure to Western scholarship has not always led to the same kind of response. On the one hand, some conceal any sense of inferiority and present themselves as specialists who write solely for a Japanese audience. At the other extreme are those who see what is really demanded of one who would work on the frontiers of research and who, therefore, abandon scholarship altogether.

Some years ago two professors, specialists in the general area of Western history and literature at a leading Japanese university, went abroad to study in the West. The first dashed off article after article upon his return with a new sense of authority. The other attempted to participate as fully as possible in the research groups and seminars of Western specialists in his field, but always wound up feeling woefully inadequate. Driven to the conclusion that he knew little more about his subject than what Western students learn in high school, he ceased to pretend that he was a scholar engaged in original research.

Let us suppose that there was a Japanese who, being an admirer of Descartes, went to Europe to study him. When he set foot on Western soil he would obviously find many scholars engaged in Cartesian studies and be confronted by heavy stacks of scholarly research that had accumulated over the centuries. He would find books written by men who had constructed their interpretations after visiting the places where Descartes lived and collecting all the relevant documents. Before our Japanese scholar came to Europe, he had assembled a collection of Descartes' original writings and, little by little, made notes of his own thoughts. Yet having seen what the long history of Cartesian scholarship has produced, he cannot but feel that this body of literature already includes views very similar to his own. Given the value modern scholarship places on originality, he decides, it would surely refuse to register the opinions of a single Japanese scholar. With this conclusion our would-be Cartesian scholar takes his place alongside many other able individuals who have begun their studies out of a genuine admiration for Western thought or Western literature and then had their spirits broken by an encounter with the modern scholarship of the West.

Miscarriages of talent are always saddening. But more is involved here than a failure of nerve. Modern scholarship itself may have to shoulder part of the blame. The normal science approach to scholarship that has undone many a Japanese student of the West looks upon research as a kind of *jindori* game in which the object is to occupy strategic pieces of territory (this approach is perhaps most common in the "field sciences," which are, as I have previously observed, typical normal sciences). Thus a certain aspect of Descartes is described as "already taken"—exhaustively researched by someone else. The assumption behind such a statement is that no matter who reads the documents, he is not likely to come up with anything new at this stage. Hence one is told that it would be useless to repeat the research and that one would be well advised (as they say) to choose another topic. The enthusiast may protest that he likes Descartes, that he "just has to" work on him, but his sentiments will not get much of a hearing from the hard-nosed professional. Research, it will be explained, is not a matter of individual likes and dislikes. In the world of modern scholarship, the true professional qualifies himself by doing some research that no one else has done, writing it up, and adding to the storehouse of knowledge. Viewed in these terms, modern scholarship seems to have very little to do with the simple joys of scholarship. It is not intellectual adventure: it is a job.

The study of Descartes may not be the best illustration of

the type of attitude I have been describing, for the philosopher can obviously be approached in a variety of ways, but in fields that have been more highly developed as normal sciences, much research work may be likened to the laying of tiles on a flat surface. In the kind of field science research where there is no need to repeat previous studies, the tiles are laid one by one until the whole pattern comes into view. This is the goal at which the positive sciences seek to arrive. A Japanese may participate in this "tile-laying" process, but the first thing he will find upon doing so is that most of the good spaces have already been taken. Of course the surface remains filled with empty spaces, and plenty of lesser places to lay a tile can still be found. In order to do so, however, the Japanese will have to work ten times harder than a Westerner if he is to overcome the handicaps of language and basic cultural background. To put it another way, with equal effort the Westerner can lay ten tiles in his lifetime to one for the Japanese. I would not by any means say that the Japanese labors in vain here, but there is a problem of efficiency.

It is true that the handicaps and barriers have been reduced in recent years. The Japanese tradition of Western studies may have started out with a group of language teachers and general interpreters of Western culture, but it has had a secure institutional base for generations and now comprises a sizable and growing community of trained scholars, some of whom presumably converse with their colleagues on occasion. With the growing use of microfilm and improved research collections, the problem of insufficient documents would also seem on its way to resolution. Moreover, as air travel grows cheaper and the yen more valuable, the researcher may be able to move around the world with such ease that local unavailability of materials will no longer be a serious inconvenience. At the same time, frequent trips abroad will make it easier for the scholar to acquire and refresh his feeling for his subject.

The West, like Japan, is a cultural admixture of historical traditions, geographical characteristics, and elements of manners and morals that do not yield easily to our understanding. With its universalistic theory, modern science is perhaps more free of these elements than any other area. On the other hand, the ancient civilizations seem alien to the contemporary Westerner as well as to the Japanese, and as time goes by Western and Japanese images of antiquity can be expected to draw closer together. When all this has been said, however, one can still not envisage a point in the normal science approach to the study of the West at which the situation will reverse itself and the Japanese scholar will have an advantage over his Western counterpart—though surely more than one Japanese scholar has struggled desperately to locate such an advantage.

The Japanese Vantage Point

Comparative history, comparative literature, comparative what-have-you—today everyone is looking for a chance to be "comparative." Sometimes the results are all too facile. This book can hardly claim to be free of this urge, though in reality it is nothing more than a "contraposition" which charts the historically divergent development of two different academic traditions. Given the nature of the subject matter, it could hardly have been otherwise. A study of the contact between different paradigms and cultures does not present significant methodological problems if the interface between them constitutes a clearly defined field. In order to compare two things that lack significant points of contact, as in the present case, one must establish some neutral coordinates. And though I believe Japan offers some advantages in this respect, it is not culturally equidistant from the two traditions under consideration. When comparing Japan and China, Japanese often remain unable to transcend a Japanese point of view. Thus I have sometimes been impressed by the way in which the judicious appraisal of a Western specialist on East Asia has exposed our blind spots.

At the same time, Westerners' treatments of their own history reflect either an effusive pride or—in reaction—a narrow parochialism, whether Westerners are conscious of it or not. Even in the world of natural sciences where there seem to be objective standards, English supporters of Newton once feuded with the continental supporters of Leibnitz, and today there are points at which Soviet and American science seem to be operating with different paradigms and values. To the extent that both are unreflectively immersed in their own habits of thought, there is room for a few fresh questions—just as someone outside a closed academic group can make positive use of his detachment.

The detachment in this case, however, is less a matter of our retreating to the sidelines than a recognition that no matter how much we might like to climb into the ring ourselves, that ring is in the West. Still, from our point of view, the fact that the West has been the main stage of modernization does not make Western studies of the West anything more than what they are—local histories employing local methodologies. This being the case, there is need for an umpire, and critics as well.

Criticism that comes from within the Western academic tradition cannot help but be parochial. The task of independent evaluation is best entrusted to a judge who stands outside the tradition. For a Western scholar to have his work translated into Japanese is now considered professionally significant. Let us gradually take up the works of Western scholars that are so much a part of our scholarly lives and write our assessment of them, preferably in a Western language. Should this happen on a significant scale, Western scholars will be forced to take note of how their work is being received in Japan, and Japan will begin to serve as a vehicle and reminder of the need for objective judgment.

But we must not stop with critical review. Japanese, it would seem, know ten times more about the West than Westerners know about Japan and the East. Indeed, in discussing world history or thought, academically backward nations have a wider field of vision. It is at this point that they can be said to stand in an advantageous position. Although there are obviously vast differences in the perspectives with which scholars in the West view the world, when compared to Western specialists in their own tradition the non-Westerners always have broader horizons.

I have on occasion been asked to serve on the editorial board of journals that bill themselves "international"—if only because a publication can no longer call itself international unless it has some people besides Europeans and Americans on its board. In these meetings I often find myself arguing for greater breadth of vision. I have also been known to suggest that the editorial operation be shifted to some place outside Europe or North America. Presumably my reasons for this are clear, but since the board members as a group really know only the West, they have not been prepared to relinquish editorial control. At the same time, the non-Western members of the board who are brought in are treated simply as specialists on the cultural area from which they come. This always makes me want to start an international magazine in Japan to set our critical vision before the world, though the fact that Japanese is not an international language makes this rather impractical.

We are living in what has been described as the age of English-language imperialism—the dominance of the English-speaking peoples in the international arena. However, the dominance of one language group or people in the history of science has usually not lasted very long. It is commonly at its peak for only one or two generations. When the center of activity does shift elsewhere, academic concepts are thrown into an alien cultural milieu—an encounter which sometimes gives rise to new breakthroughs, new revolutionary concepts. Even the most flourishing science will eventually stagnate if it submerges itself in one language and culture and loses its concern with anything outside its perimeters. One suspects, in fact, that the transplantation of paradigms from one language and culture to another has itself played a major role in the periodic revitalization of the Western academic tradition, even as it has resulted in the repeated shifts in its vital center that we have surveyed in this study. Had the center of the Western academic tradition remained in Greece, it would in all likelihood have been plagued by the same kind of stagnation that afflicted the Chinese tradition.

Our predecessors began to cultivate a critical eye for comparing the academic traditions of East and West in the Edo period when "Dutch science" was introduced on top of the centuries-old Chinese science base. Nurtured by history, this peculiar habit of mind would seem to be particularly valuable in a time like the present when many believe that the industrialization of science and learning has brought us to a critical turning point. To turn an exact and impartial eye on the world academic community in these days: this is our historical mission.

Bibliography

Abelson, Paul. *The Seven Liberal Arts: A Study in Medieval Culture*. Reprinted New York: AMS Press, 1976.

Africa, Thomas W. *Science and the State in Greece and Rome*. New York: Wiley, 1968.

Amano Ikuo. "Higher Education and Social Mobility in Modern Japan," *Journal of Educational Sociology*, no. 24, 1969.

Bacon, Francis. *The Advancement of Learning*. 1605. Available in reprint form from several publishers.

Beaujouan, Guy. "Motives and Opportunities for Science in the Medieval Universities," in Crombie, A.C. (ed.) *Scientific Change*. New York: Basic Books, 1963.

Bernal, J.D. *The World, the Flesh and the Devil: An Inquiry into the Enemies of the Rational Soul*. London: Kegan Paul, 1929.

Bernal, J.D. *Science in History* (4 vols.). London: Dent, 1957. Reprinted Cambridge, Mass.: MIT Press, 1971.

Booker, Peter Jeffrey. *A History of Engineering Drawing*. London: Northgate Publishing Company, 1963.

Brown, Harcourt. *Scientific Organizations in Seventeenth Century France, 1620–1680*. Reprinted New York: Russell and Russell, 1967.

Bunmei Genryū Sōsho [*The Sources of Civilization Library*] II, 1914.

Burnet, John. *Early Greek Philosophy*. Reprinted New York: Barnes and Noble, 1963.

Cardwell, D.S.L. "The Development of Scientific Research in the Modern University: A Comparative Study of Motives and Opportunities," in Crombie, A.C. (ed.) *Scientific Change*. New York: Basic Books, 1963.

Clavius, Christopher. *In Sphaeram Ioannis de Sacro Bosco, Commentarius*. 1607.

Clopham, Michael. "Printing," in Singer, Charles (ed.) *A History of Technology III*. Oxford: Oxford University Press, 1957.

Costello, William T. *The Scholastic Curriculum at Early Seventeenth Century Cambridge*. Cambridge, Mass.: Harvard University Press, 1958.

Crombie, A.C. *Augustine to Galieo: Science in the Late Middle Ages and Early Modern Times*. Cambridge, Mass.: Harvard University Press, 1979.

Crowther, J.G. *British Scientists of the Twentieth Century*. London: Routledge and Kegan Paul, 1952.

Drucker, Peter F. "The Technological Revolution: Notes on the Relationship of Technology, Science, and Culture," *Technology and Culture* II, 1961, pp. 342–351.

Edelstein, Ludwig. "Motives and Incentives for Science in Antiquity," in

Crombie, A.C. (ed.) *Scientific Change*. New York: Basic Books, 1963.
Edwards, Edward. *Memoirs of Libraries* (2 vols.). London, 1859.
Feuer, Lewis. *The Scientific Intellectuals*. New York: Basic Books, 1963.
Forbes, P.B.R. "Greek Pioneers in Philology and Grammar," *Classical Review*, 47, 1933, p. 105ff.
Fourcy, A. *Histore de l'Ecole Polytechnique*. Paris, 1828.
Fransson, Martha Caroline. "The Ecole Polytechnique; A study of the curriculum: 1794–1799." Harvard Honor Thesis, 1978.
Garrison, Findling H. "The Medical and Scientific Periodicals of the Seventeenth and Eighteenth Centuries," *Bulletin of the Institute of Historical Medicine,* II, 1934, pp. 285–384.
Gillispie, Charles Coulston. *The Edge of Objectivity*. Princeton, N.J.: Princeton University Press, 1960.
Goodrich, H.B., *et al.* "The Origin of U.S. Scientists," in Barber, Bernard and Hirsch, Walter (eds.) *The Sociology of Science*. Reprinted Westport, CT: Greenwood Press, 1978.
Goto Sueo. *Shina Shisō no France Seizen* [*The Advance of Chinese Thought into France*]. Tokyo: Dai-Ichi Shobo, 1933.
Graham, A.C. "China, Europe and the Origins of Modern Science," in Nakayama, S. and Sivin, N. (eds.) *Chinese Science*. Cambridge, Mass.: Harvard University Press, 1973, p. 61.
Graham, Loren. "The Formation of Soviet Research Institutes: A Combination of Revolutionary Innovation and International Borrowing," in "Texts of Symposia" for the "Professionalization of Science" Symposium at the XIVth International Congress of the History of Science, 1974.
Gutting, G. (ed.) *Paradigms and Revolutions*. Notre Dame, Ind.: University of Notre Dame Press, 1980.
Hahn, Roger. *The Anatomy of a Scientific Institution: The Paris Academy of Sciences, 1666–1803*. Berkeley: University of California Press, 1971.
Haskins, C.H. *The Rise of Universities*. Reprinted Ithaca, New York: Cornell University Press, 1957.
Hayek, F.A. *The Counter-revolution of Science. Studies in the Abuse of Reason*. Reprinted Indianapolis, Ill.: Liberty Fund, 1980.
Hiraoka Takeo. *Keisho no Seiritsu* [*The Making of the Classics*]. Osaka: Zenkoku Shobo, 1946.
Hiroshige Tetu. "Kagaku ni okeru Kindai to Gendai" ["The Modern and the Contemporary in Science"] in Satō Tomo'o (ed.) *Rekishi toshite no Gendai Shakai*. Tokyo: Chūō Daigaku Shuppanbu, 1973.
Hofstadter, Richard and Hardy, C. Dewitt. *The Development and Scope of Higher Education in the United States*. New York: Columbia University Press, 1952.
Hollinger, David H. "T.S. Kuhn's Theory of Science and Its Implications for History. *American Historical Review,* vol. 78, no. 2, April, 1973.
Hu Shih. *The Development of the Logical Method in Ancient China*. Cambridge, Mass.: Harvard University Press, 1928.
Hunt, Everett Lee. "Plato and Aristotle on Rhetoric and Rhetoricians," in Howes, Raymond (ed.) *Historical Studies of Rhetoric and Rhetoricians*. Ithaca, N.Y.: Cornell University Press, 1961, pp. 20–21.
Kaizuka Shigeki. *Kodai Chugoku no Seishin* [*The Spirit of Ancient China*]. Tokyo: Chikuma Shobo, 1967.
King, M.D. "Reason, Tradition and the Progressiveness of Science," *History*

and Theory, X, 1971, pp. 3–32.

Kneale, William and Kneale, Martha. *The Development of Logic*. Oxford: Oxford University Press, 1962.

Kōno Yoichi. *Gakumon no Magarikado* [*Learning at the Crossroads*]. Tokyo: Iwanami Shoten, 1958.

Kramer, H.J. "Arete, bei Platon und Aristoteles," *Abhandlungen d. Heidelberger Akad. d. Wiss. Philo.-hist. Kl.*, 6, 1959, pp. 380–486.

Kronhauser, W. *Scientists in Industry: Conflict and Accommodation*. Berkeley, Cal.: University of California Press, 1962.

Kronick, David A. *A History of Scientific and Technical Periodicals*. New York: Scarecrow Press, 1962.

Kuhn, T.S. *The Structure of Scientific Revolutions*, 2nd ed. Chicago: University of Chicago Press, 1970.

Lakatos, Imre and Musgrave, A. (eds.) *Criticism and the Growth of Knowledge*. Cambridge: Cambridge University Press, 1970.

Lehman, Harvey C. *Age and Achievements*. Princeton, N.J.: Princeton University Press, 1953.

Leon, Antoine. *Historie de l'Education Technique*. Paris, 1961.

Levin, S. "Socrates' Rejection of Science," *Transactions of the American Philological Association*, 79, 1948, pp. 343–344.

Li Kuo-tiao. *Chungkuo Shu-chien Pien* [*A History of Chinese Books and Documents*]. Peking, 1956. Japanese translation: Matsumi Hiromichi, *Tosho no Rekishi to Chugoku*. Tokyo: Risōsha, 1963.

Li Shu-hua. *Chungkuo yinshuashu ch'iyüan*.

Lynd, S. "Historical Past and Existential Present," in Roszak, T. (ed.) *The Dissenting Academy*. New York: Pantheon Books, 1968.

Marrou, H.I. *A History of Education in Antiquity*, trans. Lamb, George. London: Sheed and Ward, 1956.

McLuhan, Marshall. *The Gutenberg Galaxy: The Making of Typographic Man*. Toronto: University of Toronto Press, 1962.

Merton, Robert K. *Social Theory and Social Structure*. Reprinted New York: Free Press, 1968.

Mikami Yoshio. "Chugoku Sūgaku no tokushoku" ["Special Characteristics of Chinese Mathematics"], *Tōyō Gakuhō*, vol. 16, 1926, pp. 107–108.

Miyazaki Ichisada. *Kakyo* [*The Chinese Examination System*]. Osaka: Akitaya, 1946.

Morrell, Jack and Thackray, Arnold. *Gentlemen of Science: Early Years of the British Association for the Advancement of Science*. Oxford: Oxford University Press, 1981.

Nakayama Shigeru. "The Future of Research—A Call for a 'Service Science,'" *Fundamenta Scientiae*, vol. 2, no. 1, 1981, pp. 85–97.

Nakayama Shigeru. "Kindai Kaguku no Daigaku ni taisuru inpakuto II—Ekoru Poritekuniku to Kindai Kogaku no Seiritsu" ["The Impact of Modern Science on the University II: the École Polytechnique and the Establishment of Modern Engineering"], *Daigaku Ronshū*, vol. II, 1974, pp. 66–76.

Nakayama Shigeru. "Daigakushi" ["University History"], *Kagakushi Kenkyū*, No. 100, 1971, pp. 205–207.

Nakayama Shigeru. *Senseijutsu* [*Astrology*]. Tokyo: Kinokuniya Shoten, 1964.

Nakayama, S., Swain, D., and Yagi. E. (eds.) *Science and Society in Modern Japan*. Tokyo: University of Tokyo Press, 1974.

Nasr, Hossein. *Science and Civilization in Islam.* Cambridge, Mass.: Harvard University Press, 1968.

Needham, Joseph. *Science and Civilisation in China* II. Cambridge: Cambridge University Press, 1956.

Neugebauer, Otto. *The Exact Sciences in Antiquity*, 2nd ed. New York: Dover, 1969.

Nishikawa Joken. *Temmongiron* [Discourse on Astronomy]. 1712.

Ogura Kinnosuke. "Kaikyu Shakai no Sanjutsu" ["The Arithmetic of a Class Society"], *Shisō*, August and December, 1929.

Ornstein, Martha. *The Role of Scientific Societies in the Seventeenth Century.* Reprinted New York: Arno Press, 1976.

Overton, John. "A Note on Technical Advances in the Manufacture of Paper," in Singer, Charles (ed.) *History of Technology III.* Oxford: Oxford University Press, 1957.

Pan Chi-hsing. *Ching-kuo tsao-chih i-shu-shih k'ao* [*On the History of Paper-making in China*]. Peking, 1979.

Paulsen, Friederich. *The German University and University Study*, Thilly, Frank and Elwang, William W. (trans.). New York: C. Scribner's Sons, 1906.

Periphanakis, Constantin. *Le Sophistes et le Droit.* Athens, 1953.

Pfeiffer, Rudolf. *History of Classical Scholarship from 1300 to 1850.* Oxford: Oxford University Press, 1968.

Price, Derek J. de S. "*Communication in Science—Philosophy and Forecast*," in De Reuck, Anthony and Knight, J. (eds.) *Communication in Science.* London: Churchill, 1967.

Purver, Margery. *The Royal Society; concept and creation.* London: Routledge and Kegan Paul, 1967.

Rashdall, Hastings. *The Universities of Europe in the Middle Ages,* vol. 1. Powicke, F.M.
and Emden, A.B., eds. Oxford: Oxford University Press, 1936.

Ravetz, Jerome R. *Scientific Knowledge and Its Social Problems,* Part II. Oxford: Oxford University Press, 1971.

Rosenthal, Fr. *The Technique and Approach of Muslim Scholarship.* Rome, 1947.

Russell, Bertrand. *A History of Western Philosophy.* New York: Simon and Schuster, 1945.

Sayili, Aydin. "Higher Education in Medieval Islam," *Annales de l'Universite d'Ankara,* II, 1948, pp. 64–65.

Sayili, Aydin. *The Observatory in Islam.* Ankara, 1960.

Shimada Yūjiro. *Europpa Daigakushi Kenkyū* [Studies in the History of the European University]. Tokyo: Miraisha, 1967.

Shinjō Shinzō. *Tōyō Temmongakushi Kenkyū* [Studies in the History of East Asian Astronomy]. Kyoto: Kōbundō, 1928.

Solmsen, Friedlich. "Aristotelian Tradition in Ancient Rhetoric," *American Journal of Philology* 62, 1941, p. 35ff.

Sprat, Thomas. *The History of the Royal Society of London.* Reprinted Seattle: University of Washington Press, 1966.

Stimson, Dorothy. "Puritanism and the New Philosophy in Seventeenth Century England," *Bulletin of the Institute of Historical Medicine,* III, 1935, p. 321.

Struever, Nancy S. *The Language of History in the Renaissance.* Princeton, N.J.: Princeton University Press, 1970.

Sūgakushi Kenkyū [*Studies in the History of Mathematics*], no. 2, Tokyo, 1948.

Sugimoto Isao. *Kindai Jitsugakushi no Kenkyū*. Tokyo: Yoshikawa Kōbundō, 1962.

Taga Akigoro. *Tōdai Kyōikushi no Kenkyū* [Studies in T'ang Educational History]. Tokyo: Fumaidō, 1953.

Teng Ssu-yü. "Chinese Influence on Western Examination Systems," *Harvard Journal of Asiatic Studies*, VII, 1943, p. 227.

Tillyard, A.I. *A History of University Reform from 1800 A.D. to the Present Time*. Cambridge: Cambridge University Press, 1913.

Tōyama Hiraku. "Noryoku to Shiken to Gakko to" ["Ability, Examinations, and Schools"], *Sekai*, November 1971.

Tsuda Sōkichi. *Rongo to Koshi no Shisō* [The 'Analects' and the Thought of Confucius]. Tokyo: Iwanami Shoten, 1948.

Tsuda Sōkichi. "Jukyō Seiritsushi no Ichisokumen" ["One Aspect of the Establishment of Confucianism"], collected in *Tsuda Sōkichi Zenshū*, vol. 16. Tokyo: Iwanami Shoten, 1965, p. 56.

Veblen, Thorstein. *The Higher Learning in America; a memorandum on the conduct of universities by businessmen*. 1918. Reprinted New York: Augustus M. Kelley Publishers, 1973.

Watson, Burton (trans.) *Records of the Grand Historian of China* (2 vols). New York: Columbia University Press, 1961.

Wherli, F. "Aristoteles in der Sichtseiner Schule, Platonisches und Vorplatonisches," *Aristotle et les problems de methode*. Louvain, 1960, p. 336.

Index

INDEX 281